CHEMICAL BONDS IN SOLIDS

Volume 3: X-Ray and
Thermodynamic Investigations

CHEMICAL BONDS IN SOLIDS

Volume 1: General Problems and Electron Structure of Crystals
Volume 2: Crystal Structure, Lattice Properties, and Chemical Bonds
Volume 3: X-Ray and Thermodynamic Investigations
Volume 4: Semiconductor Crystals, Glasses, and Liquids

CHEMICAL BONDS IN SOLIDS

Proceedings of the International Symposium on Chemical Bonds in
Semiconducting Crystals held in Minsk, USSR, in 1967

Edited by
Academician N. N. Sirota
Institute of Solids and Semiconductors
Academy of Sciences of the Belorussian SSR, Minsk

Translated by
Albin Tybulewicz
Editor, *Soviet Physics — Semiconductors*

Volume 3: X-Ray and
Thermodynamic Investigations

ⓒ⁄ⓑ CONSULTANTS BUREAU • NEW YORK–LONDON • 1972

This series of four volumes is comprised of articles appearing in *Khimicheskaya Svyaz' v Kristallakh* and *Khimicheskaya Svyaz' v Poluprovodnikakh* (Minsk: Nauka i Tekhnika, 1969). The articles are arranged topically and have been revised and corrected by the editor for this edition. The exact source of each article appears in a footnote on the opening page of that article.

Many of the articles by non-Russian authors are printed from manuscripts kindly furnished by the authors.

The present translation is published under an agreement with Mezhdunarodnaya Kniga, the Soviet book export agency.

Н. Н. СИРОТА

ХИМИЧЕСКАЯ СВЯЗЬ В КРИСТАЛЛАХ
KHIMICHESKAYA SVYAZ' V KRISTALLAKH

ХИМИЧЕСКАЯ СВЯЗЬ В ПОЛУПРОВОДНИКАХ
KHIMICHESKAYA SVYAZ' V POLUPROVODNIKAKH

Library of Congress Catalog Card Number 73-185456

ISBN-13: 978-1-4684-1688-6   e-ISBN-13: 978-1-4684-1686-2
DOI: 10.1007/978-1-4684-1686-2

© 1972 Consultants Bureau, New York
Softcover reprint of the hardcover 1st edition 1972
A Division of Plenum Publishing Corporation
227 West 17th Street, New York, N.Y. 10011

United Kingdom edition published by Consultants Bureau, London
A Division of Plenum Publishing Company, Ltd.
Davis House (4th Floor), 8 Scrubs Lane, Harlesden, London,
NW10 6SE, England

PREFACE

The present four volumes, published under the collective title of "Chemical Bonds in Solids," are the translation of the two Russian books "Chemical Bonds in Crystals" and "Chemical Bonds in Semiconductors." These contain the papers presented at the Conference on Chemical Bonds held in Minsk between May 28 and June 3, 1967, together with a few other papers (denoted by an asterisk) which have been specially incorporated. Earlier collections (also published by the Nauka i Tekhnika Press of the Belorussian Academy of Sciences) were entitled "Chemical Bonds in Semiconductors and Solids" (1965) and "Chemical Bonds in Semiconductors and Thermodynamics" (1966) and are available in English editions from Consultants Bureau, New York (published in 1967 and 1968, respectively).

The subject of chemical bonds in crystals, including semiconductors, has recently become highly topical and has attracted the interest of a wide circle of physicists, chemists, and engineers.

Until recently, the most successful description of the properties of solids (including semiconductors) has been provided by the band theory, which still dominates the physics of solids. Nevertheless, it is clear that the most universal approach is that based on the general theory of chemical bonds in crystals, in which details of the electron distributions between atoms and of the wave functions appear quite explicitly.

Although, in principle, the general theory is superior to the band theory, the appropriate techniques for its application are not yet developed sufficiently well and a unified approach to a quantitative description of the structures and the physical properties of crystals is still lacking. The less generally valid band theory can at present give clearer and more convincing explanations of changes in the physical properties of crystals caused by variations in the temperature, pressure, magnetic and electric fields intensities, impurity concentrations, etc. However, many problems encountered in the study of chemical bonds in crystals cannot be considered within the framework of the standard band theory. They include, for example, determination of the elastic, thermal, and thermodynamic properties of solids, as well as the structure and properties of liquid and amorphous semiconductors.

Among the most important requirements in the theory of chemical bonds is the development of a unified method for the description of the chemical interaction between atoms, which would be based on the structure of the atomic electron shells and in which one would utilize the wave functions and the electron density distributions calculated for isolated (free) ions on the basis of the data contained in Mendeleev's periodic table of elements. This unified approach should make it possible to elucidate the interrelationship between the various physical properties and the relationship between the equilibrium and the excited energy states in crystals. In contrast to the study of chemical bonds in a molecule, an analysis of the atomic interaction in crystals must make allowances for the presence of many coordination spheres, the long- and short-range symmetry, the long- and short-range order, and other special features of large crystalline ensembles. As mentioned already, the band theory is intimately related to the chemical interaction between atoms. An analysis of the published experimental and theoretical data shows that the need is urgent, and all the prerequisites now exist, for the development of the proposed unified theory of chemical bonds, which would include, as one of its essential components, an improved band theory following logically from the general concept of chemical bonds in crystals.

Various approaches are possible to the theoretical solution of the problem of chemical bonding. One can solve the problem by the purely theoretical technique of finding wave functions and using quantum-mechanical methods without recourse to any additional empirical constants. Slater pointed out that this is the most direct way although, obviously, it is not the fastest or the most effective. This purely theoretical approach requires the knowledge of wave functions to a high degree of precision. The existing quantum-mechanical methods are approximate. Solutions of the Schrödinger equation for atomic systems depend on the reduction of the problem to the one-electron approximation or, in the case of the many-electron approach, on approximations of varying degrees of precision. These two approaches yield relatively accurate results only if the problem is tackled scrupulously and a very careful allowance is made for the errors committed. In calculations of the energy relationships and of the lengths of atomic bonds in crystals, the required parameters are small differences between large quantities, and therefore considerable errors may be committed if one does not properly control the approximations made.

The established methods of valence bonds and molecular orbitals (MO), including the method of linear combinations of atomic orbitals (LCAO), which have been so successful in the treatment of molecular systems, need further refinement when applied to crystals. The preference for the method of valence bonds in the case of solids is not accidental because it yields clearer results. In contrast, calculations dealing with the simplest molecules are currently tackled usually by the method of molecular orbitals, including the method of absolute, purely theoretical, quantum-mechanical calculations, which is adopted in those cases when a sufficiently precise form of the Hamiltonian operator can be obtained for the system being considered.

Various semiempirical methods of solving the quantum-mechanical problem of the interaction between atoms in a crystal are also of great interest. In these methods, the atomic distances, energies, and interactions are calculated by invoking the mathematical apparatus of quantum mechanics in conjunction with empirical or semiempirical wave functions (including those found experimentally). In spite of its well-known inaccuracies and inconsistencies, the semiempirical method for the quantum-mechanical solution of the problem of chemical bonding in crystals provides the most accurate results for a given amount of work.

The use of experimentally determined wave functions may provide the most effective technique among the semiempirical methods. One way of developing these methods would be to use the statistical approach to chemical bonding in crystals, based on the statistical theory of atoms developed originally by Thomas, Fermi, and Dirac, and more recently by Gombás. The statistical methods have the advantages of simplicity and clarity but they are the least accurate. However, the statistical theory of atoms has recently been refined so that the accuracy of the results obtained is now higher. By way of example, we can cite the book by Gombás which gives the values of the wave functions (atomic orbitals) for all elements in the periodic system. These functions are calculated according to the statistical model of the atom, and are in good agreement with the results of the more complex and more laborious Hartree−Fock calculations.

In many quantum-mechanical calculations, use is made of the wave functions obtained by the Dirac−Slater and the Hartree−Fock methods for the approximate solution of the Schrödinger equation for free atoms. It would be very interesting to determine whether these functions could be refined specifically for crystals and whether the problem could be solved using relatively simple analytic approximations to the calculated functions. In particular, the approximation by Gaussian functions demands attention.

It would also be interesting to examine the possibility of using directly, in quantum-mechanical solutions of problems in the quantum chemistry of crystals, the tabulated data on the wave functions and atomic scattering factors calculated by the Hartree—Fock and Dirac—Slater methods for ions with different degrees of ionization. When the crystal lattice of an element or a compound is formed, the wave functions undergo a slight change which can be allowed for. Therefore, the solution of problems in the quantum chemistry of crystals through the use of the tabulated wave functions of ions with different degrees of ionization, combined with allowance for the crystal field, has definite advantages and provides a promising approach. In particular, the tabulated data can be used to estimate approximately the ionic radii, the band structure, and the bond energies if an allowance is made for the changes introduced by the crystal field.

It follows that, in all these cases, it is desirable and even essential to have sufficiently accurate experimental data which can be used as the basis of comparison with the theoretical values. Thus, the establishment of the experimental quantum chemistry of crystals was a great step forward in the development of a unified quantum-chemical theory of bonds in crystals.

The Minsk conferences on chemical bonds in semiconductor and other crystals have demonstrated clearly the importance of experimental determinations of the distribution of the electron density in crystals, the distribution of the potential in the crystal lattice, and the application of various methods to the calculation of the effective charges of ions and of accurate values of the atomic spacings and bond energies. It has been found possible to estimate various physical properties of crystals from the experimentally and theoretically determined atomic scattering functions and the electron density distributions in crystals. These problems are considered in several papers in the present collection.

The books we are presenting deal with various aspects of chemical bonding in crystals, particularly in semiconductor crystals. In semiconductors, the nature of the chemical bonds can vary within wide limits and extend over a considerable part of the well-known tetrahedron of the metallic, covalent, ionic, and van der Waals types of chemical bond. This tetrahedron is only approximate but it provides a sufficiently clear basis for the classification of transitions which may occur between different types of bond.

Some workers regard the atomic bonds in semiconductors as a special type which is independent of the other four types. It seems clear, however, that the chemical bonds in semiconductors do not represent an independent type of bonding but form the most general class, which extends from the center of the aforementioned tetrahedron to its vertices.

The electron distributions in the atoms forming a crystal, the dimensions of the ions, the ionization potentials, and the type and the energy of the atomic bonds in the crystal, all depend on the positions in Mendeleev's table of the elements forming the crystal. Consequently, an analysis of the changes in the nature of chemical bonds resulting from changes of the positions of a crystal's components in Mendeleev's table provides a convincing and clear picture of the sequence of changes in the nature and the energy of the interaction between atoms.

A characteristic feature of the present state of the science of chemical bonds in crystals is the tendency to combine theoretical and experimental investigations and to close the usual gap between theoretical analyses and experimental studies.

Recent years have seen a considerable extension of the experimental methods used in quantum chemistry and in investigations of the nature of chemical bonds in crystals. It is worth mentioning methods based on the studies of the elastic and the inelastic scattering (by crystals) of x rays, electrons, neutrons, protons, mesons, α and other particles, as well as the x ray spectroscopic methods. Methods based on the use of positron annihilation are also of considerable interest.

Practically inexhaustible possibilities are latent in the methods based on the external and internal photoelectric effects excited by x rays and light of various wavelengths. It is worth mentioning specially the methods of electron (β) spectroscopy, cold emission, photoelectric emission, and photoelectron spectroscopy, the last being used widely for the purpose of chemical analysis.

Investigations of the scattering of x rays, electrons, mesons, and neutrons, carried out under suitable conditions, can provide highly accurate information on the atomic scattering functions, and on the electron and spin densities in crystals.

Improved x-ray spectroscopic methods can be used to find the effective charges of ions and to determine the spectra of the energy states N(E) of the electrons in crystals. These spectra govern many physical properties. Experimental investigations of the dependences of the N(E) spectra of the components (elements and compounds) of crystals on the positions of the elements concerned in Mendeleev's periodic table can give extensive information on some features of chemical bonding.

The density of states N(E), which governs the nature of atomic bonds, is one of the important criteria that determine the transition to the superconducting state in the theories of Bardeen, Copper, Schrieffer, and Bogolyubov. The density of states N(E) can be found directly from the results of x-ray spectroscopic analyses.

The problem of chemical bonding is inseparable from the magnetic properties of crystals. This follows clearly from the work of Klemm, Dorfman, Goodenough, and others who have found that magnetic measurements can provide quantitative information on the type and energy of atomic bonds.

There is little doubt that direct experimental determination of the heats of atomization, heats of formation of compounds, free energies, and other thermodynamic properties of crystals can provide the basis for theoretical calculations of the bond energies, which are used to establish the nature of bonds.

Other important experimental methods for elucidating the nature of chemical bonds in crystals are the study of the elastic and thermal properties and the determination of phonon spectra. Many of these properties can be regarded as thermodynamic stability criteria and are quantitative measures of the second derivatives of the bond energies with respect to the atomic spacing. Moreover, such investigations yield temperature dependences of the characteristic themodynamic functions.

The cardinal feature of successful investigations of chemical bonds in crystals is the proper combination of theoretical, experimental, and crystallochemical techniques. The crystallochemical aspect of chemical bonding has been stressed particularly clearly by Pauling, Belov, Geller, Goodenough, and others. This aspect is also reflected in the present collection.

The number of experimental physical methods which can provide quantitative information of direct interest in the quantum chemistry of crystals is increasing continuously. Routine use is made of various optical and spectroscopic methods in the investigations of solids. Gamma resonance (Mössbauer effect), Kikoin—Noskov photoelectric and photomagnetic effects, cyclotron resonance, EPR, NMR, and other methods are used widely. Measurements of mechanical properties, whose importance was stressed by A. F. Ioffe, are being used for quantum-chemical purposes but not sufficiently intensively. Many of these methods have been introduced, reached sufficient precision, and yielded satisfactory interpretations only in the last few years. Clearly, the extension of experimental investigations of the mechanics, nature, special features, and energies of chemical bonds in crystals has become a very urgent task.

The guiding principles in these investigations are those enunciated by P. L. Kapitza, who suggested that substances be investigated in extreme or limiting states such as ultra-high purity, very low temperatures, strong magnetic and electric fields, and high pressures. To these must be added the principles of physiochemical analysis proposed by N. S. Kurnakov, who suggested that investigations be carried out by varying the composition and the external equilibrium parameters.

Our four volumes reflect this tendency to combine theoretical and experimental methods in studies of chemical bonds in crystals.

These volumes contain theoretical and experimental papers on chemical bonds in crystals, especially those in semiconductors and semimetals.

The first volume deals with the general aspects of chemical bonding in crystals and with the interrelationship between the electron structure of crystals and their physical properties.

Some of the papers comprising the first volume deal with those theoretical and experimental aspects of chemical bonds in crystals which relate to the most general rules governing the dependence of the atomic interactions on the positions of the components in Mendeleev's periodic table. The criteria governing the transitions from the metallic to the semiconducting and the superconducting states are also considered.

Professor Sir Nevill Mott deals with the important topic of developing a satisfactory theory of liquid and amorphous semiconductors. Sir Nevill's paper is of great interest because of the heavy stress he lays on the importance of knowing the nature of chemical bonds in liquid and amorphous semiconductors in order to gain an understanding of the special features of the state and properties of these substances.

The first volume also includes papers on the correlation between the magnetic properties, the structure, and the electron distributions in crystals, and on that between the electron interaction, the distributions of the electron densities and potentials, and the band structure. New experimental results are also reported.

The second and the third volumes deal with the correlation between the nature of chemical bonds and the physical properties of crystals, particularly lattice dynamics, and thermodynamic and thermochemical parameters. The second volume is concerned mainly with crystal structure, physical properties, and lattice dynamics. The first part of the third volume reports extensive data on electron distributions and on the effective charges of ions deduced from x-ray diffraction and spectroscopic investigations. The second part of the third volume is concerned mainly with thermodynamic and thermochemical investigations of semiconductor crystals. This part includes also papers concerned with the thermodynamic stability of crystals.

The fourth volume concentrates mainly on the properties of semiconducting compounds, including transition-metal silicides, amorphous and liquid semiconductors, and particularly the nature of the chemical bonds in these materials.

One of the chapters in the fourth volume is devoted primarily to complex semiconducting compounds. Most of the papers comprising this chapter were contributed by the talented scientist N. A. Goryunova (who died prematurely) and her colleagues. In the first paper of this chapter, consideration is given to the methodological aspects of the search for compounds exhibiting semiconducting properties on the basis of Mendeleev's periodic table.

The four volumes are packed with information. Some of the papers are tentative and suggest possible ways of solving problems. The majority give the results of completed investigations. The whole set provides an idea of the present state of the science of chemical bonds in crystals, fills some gaps in our knowledge of the chemical bonding in condensed phases, and should stimulate further studies of this very interesting subject. It is hoped that the wide range of readers engaged in the physics, chemistry, and technology of solids and semiconductors will find something of interest to them in these volumes.

CHEMICAL BONDS IN SOLIDS, VOL. 3
N. N. Sirota

CONTENTS

PART I

X-RAY DIFFRACTION AND SPECTROSCOPIC INVESTIGATIONS

Distribution of Valence Electrons in Magnesium Silicide. 3
  D. Panke and E. Wölfel

Experimental Determination of the Structure Amplitudes and of the Degree of
Ionization in Gallium Phosphide 9
  N. N. Sirota, A. U. Sheleg, and E. M. Gololobov

Mean-Square Dynamic Displacements and Atomic Scattering Factors of Ions in
Aluminum Nitride 14
  N. N. Sirota, A. I. Olekhnovich, and N. M. Olekhnovich

X-Ray Spectroscopic Investigation of Chemical Bonding in Some Rare-Earth
Titanates 21
  A. Meisel and G. Leonhardt

Determination of the Degree of Ionicity of the Bonds in Some Compounds of
Phosphorus from the Shift of the $K_{\alpha_{1,2}}$ X-Ray Lines. 25
  É. P. Domashevskaya, Ya. A. Ugai, and O. Ya. Gukov

Derivation of the Electronic Structure of Aluminum Atoms in Al_2O_3 from X-Ray
Spectroscopic Data. 30
  S. M. Karal'nik and É. P. Domashevskaya

Investigation of the Atomic Binding Forces in Mercury Telluride by the X-Ray
Diffraction Method. 34
  L. I. Kleshchinskii, P. V. Sharavskii, and A. I. Inyutkin

Influence of the Chemical Bonding on the $K_{\alpha_{1,2}}$ Doublet of Iron Investigated
Using Local X-Ray Spectroscopic Analysis. 39
  G. N. Ronami, A. K. Milai, and L. I. Tutova

Investigation of the Carbon K and Metal Emission Bands and Bonding for
Stoichiometric and Nonstoichiometric Carbides. 42
  J. E. Holliday

Investigation of the Distribution of the Potential in the Unit Cell of Lithium
Fluoride. 58
  L. V. Shevtsov and V. P. Tsvetkov

Chemical and Knight Shifts in $A^{III}B^{V}$ Compounds. 66
  A. Lösche and S. Grande

X-Ray Emission K Spectra of Silicon in Chromium Silicides   70
    V. P. Tsvetkov and N. D. Savchenko

Electron Diffraction Study of Sodium Bromide Thin Films.   73
    A. G. Buntar' and A. F. Margolina

Anisotropy of the Scattering of X Rays and Electrons by Cubic Crystals   80
    A. G. Buntar'

Relationship Between the Reflection of X Rays and the Degree of Perfection of
Crystals in Focusing Monochromators.   87
    N. M. Olekhnovich

Electron Density Distribution and Effective Charges in Indium Phosphide   92
    N. N. Sirota, E. M. Gololobov, and A. U. Sheleg

Distribution of the Uncompensated Electron Density in the Ferromagnetic
Manganese Compounds MnAs, MnSb, and MnBi   96
    N. N. Sirota, and É. A. Vasil'ev

Distribution of the Potential in the Gallium Phosphide Lattice   101
    N. N. Sirota, A. U. Sheleg, and Zh. A. Matskevich

PART II

THERMODYNAMIC AND THERMOCHEMICAL INVESTIGATIONS

Phase Equilibria and Transitions at Low Temperatures   109
    V. K. Semenchenko

The Average Heat of Atomization and the Properties of Semiconductors   114
    V. Sadagopan and H. C. Gatos

Heats of Atomization and Some Thermochemical Constants of $A^{III}B^{V}$ Compounds   124
    L. I. Marina, A. Ya. Nashel'skii, and B. A. Sakharov

Thermodynamic Properties of Alloys of the $ZnAs_2 - CdAs_2$ and $Zn_3As_2 - Cd_3As_2$
Quasibinary Systems   130
    N. N. Sirota and É. M. Smolyarenko

Thermodynamic Properties of Gallium Phosphide   134
    A. S. Abbasov, K. N. Mamedov, and D. M. Suleimanov

Thermodynamic Properties of Manganese Germanides.   137
    V. N. Eremenko, G. M. Lukashenko, and R. I. Polotskaya

Some Relationships Governing the Doping of Semiconductors   142
    S. A. Semenkovich and Yu. P. Shishkin

Investigation of the Evaporation of Antimony and Bismuth Tellurides and of
Bismuth Selenide   151
    Z. Boncheva-Mladenova, A. S. Pashinkin, and A. V. Novoselova

Thermodynamic Properties of Bi_2Te_3, Bi_2Se_3, Sb_2Te_3, and Sb_2Se_3   159
    S. A. Semenkovich and B. T. Melekh

Enthalpies of Formation of Some Rare-Earth Sulfides   163
    S. A. Semenkovich, V. M. Sergeeva, and A. D. Finogenov

Thermodynamic Properties of $CoTe_2$.   171
    V. A. Geiderikh, Ya. I. Gerasimov, and O. B. Matlasevich

CONTENTS   xiii

Thermodynamic Properties of Crystals in Relation to the Nature and Energy of
the Interatomic Interaction   175
    N. N. Sirota

Thermodynamic Properties of AlSb and AlSb−GaSb Solid Solutions.   184
    V. V. Samokhval, A. A. Vecher, and E. P. Pan'ko

EMF Study of the Thermodynamic Properties of Gallium Selenide Ga_2Se_3.   188
    A. S. Abbasov, K. N. Mamedov, P. G. Rustamov, and P. K. Babaeva

Thermodynamic Properties of Some Semiconductors   191
    P. G. Maslov and Yu. P. Maslov

Kinetics of the Etching of A^{IV} and $A^{III}B^{V}$ Semiconductors.   196
    G. M. Orlova

PART I

X-RAY DIFFRACTION AND
SPECTROSCOPIC INVESTIGATIONS

DISTRIBUTION OF VALENCE ELECTRONS
IN MAGNESIUM SILICIDE*

D. Panke and E. Wölfel

Lehrstuhl für Strukturforschung
Eduard Zintl-Institut für anorganische und physikalische Chemie
Technische Hochschule, Darmstadt, Germany

An experimental investigation was made of the electron density distribution in various planes (xy0, xy$\frac{1}{4}$, xxz) of the Mg_2Si lattice. The experimental values of the density were compared with the values calculated from the theoretical f curves. It was established that an electron bridge representing 0.2 electron/Å^3 appears between the magnesium and silicon ions in the xxz plane, which indicates a covalent contribution to the bond. The charge distribution near the magnesium ions is approximately spherically symmetrical whereas the outer electrons of the silicon ions are displaced in the direction of the nearest magnesium ions. The silicon ions carry approximately one negative unit charge whereas the magnesium ions carry 1.5 positive charges each. The other two valence electrons of the Mg_2Si formula unit are distributed equally between these ions. The ionic radii found in this study lie between Pauling's ionic radii calculated for Mg^{2+} and Si^{4-} and the covalent radii, and are thus consistent with the calculated covalent contribution to the bonding.

A group of research workers in the Darmstadt Technische Hochschule have been studying the distribution of the valence electrons in solids since 1948, this work constituting a continuation of earlier investigations by Brill, Grimm, Hermann, and Peters [1]. The results indicated that, in principle, the electron density distribution can be used to determine unambiguously the characteristic type of chemical bond in the material under consideration.

It was considered desirable to supplement the existing data relating to alkali halides (sodium chloride [2] and lithium fluoride [3]), fluospar [4], aluminum [5], silicon [6], and diamond [7] by a corresponding analysis of Mg_2Si, chiefly in order to be able to compare the charge distribution on the silicon atoms in Mg_2Si with that in elemental silicon and hence draw appropriate conclusions regarding the corresponding degree of ionization.

---

* The articles in this volume were originally published in two Russian books, "Khimicheskaya Svyaz' v Kristallakh," published by Nauka i Tekhnika, Minsk, 1969 (hereafter called "Crystals"), and "Khimicheskaya Svyaz' v Poluprovodnikakh," also published by Nauka i Tekhnika, Minsk, 1969 (hereafter called "Semiconductors"). The source of this article is "Crystals," pp. 128-135.

Magnesium silicide occurs in the antifluorite structure and belongs to the group of Zintl phases, named after Edward Zintl [8]. He found that the elements belonging to the four groups preceding the inert gases (elements capable of forming anions) can only form valence compounds with magnesium and the structures of these compounds are similar to those of purely nonmetallic compounds. On the other hand, elements belonging to groups I and III do not form valence magnesium compounds and their structures resemble those of alloys.

If we agree with Zintl in treating Mg_2Si as a salt, then we may clearly assume an ionic structure, in which the magnesium appears as Mg^{2+} and the silicon as Si^{4-}. If we take Pauling's ionic radii (Mg^{2+}: 0.65 Å; Si^{4-}: 2.71 Å), the ionic spheres will clearly overlap because the distance between two lattice constituents calculated from the lattice constant of $a = 6.351$ Å is then 2.74 Å. However, if we use Pauling's covalent radii, which are generally valid for homopolar bonds of a tetrahedral nature (Mg: 1.40 Å; Si: 1.17 Å), the measured distance agrees far better with theory.

Welker accordingly postulated a homopolar type of bond [9]. Following Welker, we may consider that only four out of the eight valence directions of the silicon atom are operative at any particular instant. They are directed toward the vertices of a tetrahedron in accordance with the normal hybrid function and are each occupied by two electrons with antiparallel spins. The other four valences remain free. The magnesium atoms not held together by homopolar bonds thus exist as Mg^{2+} ions and are bound to the cores by means of ionic bonds; they are situated in the voids of a negatively charged zinc-blende lattice. In a second possible form of the system, the free and occupied valences are interchanged. The whole system thus constitutes the overlapping of two degenerate states, expressed in the following manner by Krebs [10]:

$$[Mg - Si]^{2-} Mg^{2+} \leftrightarrow Mg^{2+} [Si - Mg]^{2-}.$$

By virtue of his hypothesis regarding the directionally degenerate sp^3 hybrid associated with the silicon atom, Welker was able to explain the structural properties of Mg_2Si and its higher homologs (up to Mg_2Pb), as well as the semiconducting properties reported by Busch and Winkler [11].

There have been two investigations involving the x-ray determination of the electron density distribution in Mg_2Si.

Ageev and Guseva [12] used filtered copper radiation in order to determine the x-ray diffraction intensities of powder samples by the photographic technique; they deduced homopolar bonding between the ions, an absence of conduction electrons, and an asymmetrical electron density distribution around the magnesium atoms, in contrast to the silicon atoms, which had a spherically symmetrical charge distribution.

Wagner [13] tried to determine the magnitudes of the structure factors from single-crystal measurements alone; he was unable to effect a Fourier synthesis, however, owing to the severe fluctuations in the results so obtained, these being attributable partly to the difficulties encountered in preparing a suitable single crystal and partly to the strong extinction in Mg_2Si. In an earlier paper Panke [14] described a procedure for growing Mg_2Si single crystals and gave an approximate estimate of the corresponding extinctions, thus justifying a combined analysis of the electron density distribution of Mg_2Si by both single-crystal and powder measurements.

In the present investigation an Mg_2Si single crystal was prepared in the form of a plane-parallel plate, and the integral intensities of the x-ray reflections were determined in absolute units up to the order of 880 by the ionization-chamber transmission method. The secondary-extinction correction was estimated by making measurements with plates of different thicknesses. In order to correct for the very strong primary extinction, the strongest reflections were also measured for powder samples.

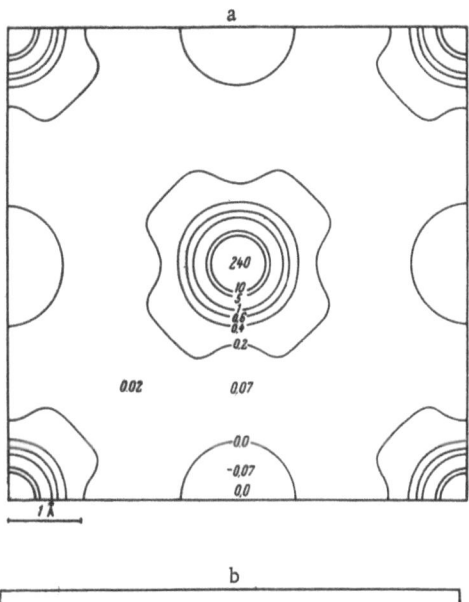

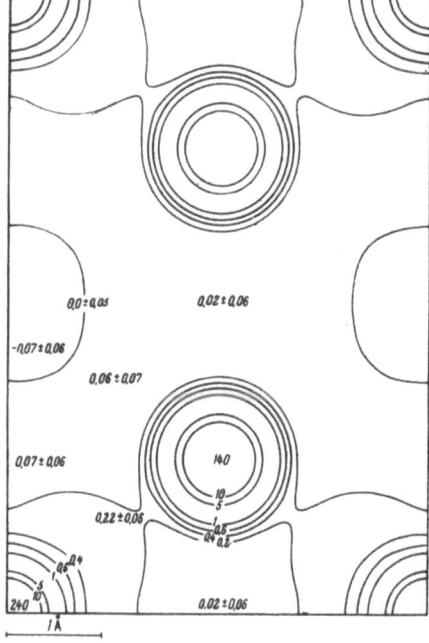

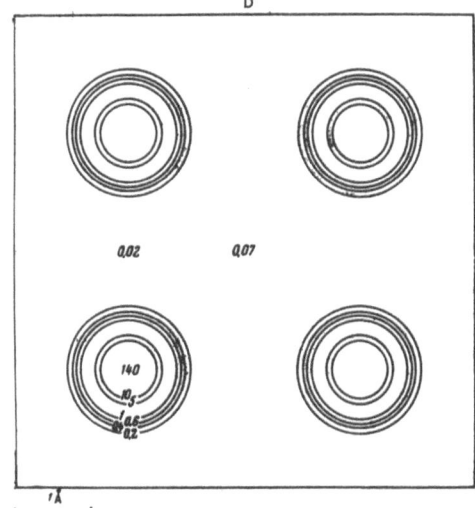

Fig. 1. Electron density distributions in various Mg$_2$Si planes (electrons/Å^3): a) xy0; b) xy$^1/_4$; c) xxz.

The structure factors derived from the measurements differed substantially from the theoretical values for the (111), (222), (331), and (422) lines. The electron density distributions calculated from the structure factors by a three-dimensional Fourier synthesis for the xy0, xy$^1/_4$, and xxz planes are shown in Fig. 1a-c.

An examination of the density values in all three planes of the Mg$_2$Si lattice shows that the density never quite vanishes at any point of the system. The only exception is the region around the $^1/_2$00 point (and cubically symmetrical equivalent points), in which the density becomes slightly negative (−0.07 electron/Å^3). The superposition of measuring errors plays a major role at such highly symmetrical points.

The average density in the space between the ions approximately equals 0.05 electron/Å^3.

A graphical construction of the electron density distribution suggests that the magnesium ions are spherically symmetrical up to a density of 0.2 electron/Å^3, whereas the outer region of the electron shell of the silicon ion is distorted in the direction of its nearest silicon neighbors.

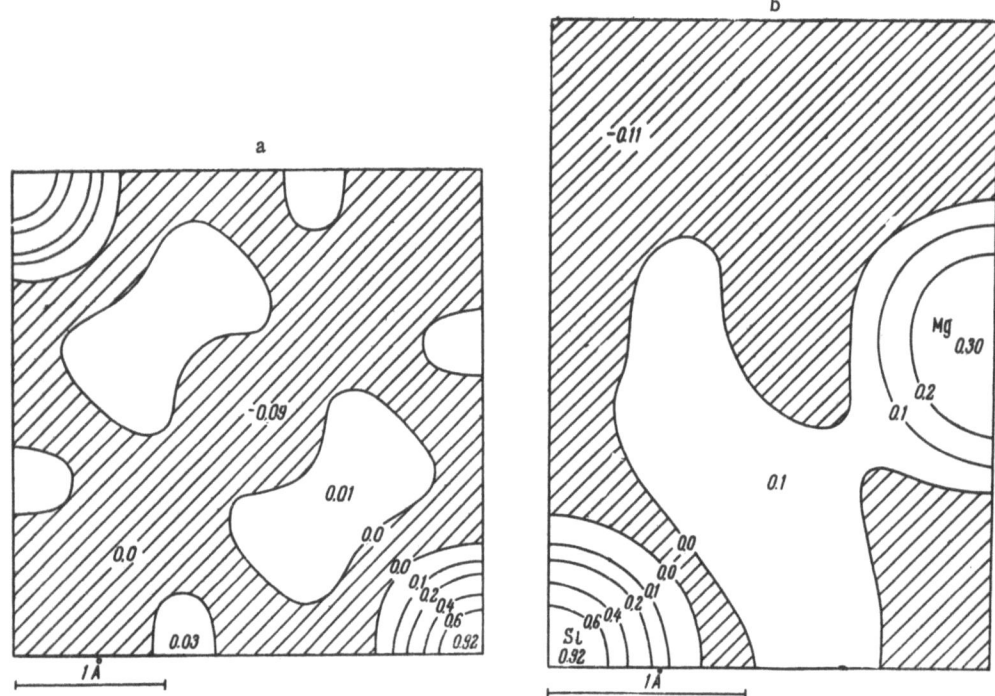

Fig. 2. Electron-density differences for various Mg_2Si planes: a) xy0; b) xxz.

In the **xxz** plane the magnesium ions attract the outer shell of the silicon ions so much that an electron bridge of 0.2 electron/ $Å^3$ develops.

In order to discuss the experimental electron density distribution of Mg_2Si, we should consider the theoretical density distribution based on the spherically symmetrical charge cloud of a free atom.

In order to facilitate comparison, the theoretical atomic form factors f_0 for the uncharged Si and Mg were converted to the values corresponding to the actual temperature during the measurements. This was done by means of the measured temperature factors $B_{Si} = 0.548$ $Å^2$ and $B_{Mg} = 0.764$ $Å^2$. The theoretical density distribution ρ_{th} was then calculated from the resultant theoretical structure factors F_{th}.

Figure 2 shows the density difference $\rho(x, y, z) - \rho_{th}$ for the xy0 and **xxz** planes. The density difference is negative in the shaded regions; this is generally the case in the region between the ions, i.e., the experimental density in these regions is lower than the density obtained from the overlap of the spherically symmetrical theoretical charge clouds. However, on the line between two adjacent silicon atoms in the xy0 plane and between the silicon and magnesium atoms in the **xxz** plane, an increase in charge is observed. At the points $1/8 \, 1/8 \, 1/8$ the density difference rises to 0.1 electron/ $Å^3$.

Figure 3 shows the variation in electron density in the [111] direction, on the line joining a silicon ion to the nearest neighboring magnesium ion (curve 1). For comparison, the fall-off in density on moving away from the silicon and magnesium ions in the [100] direction is also shown (curves 3 and 4), together with the theoretical density (curve 2).

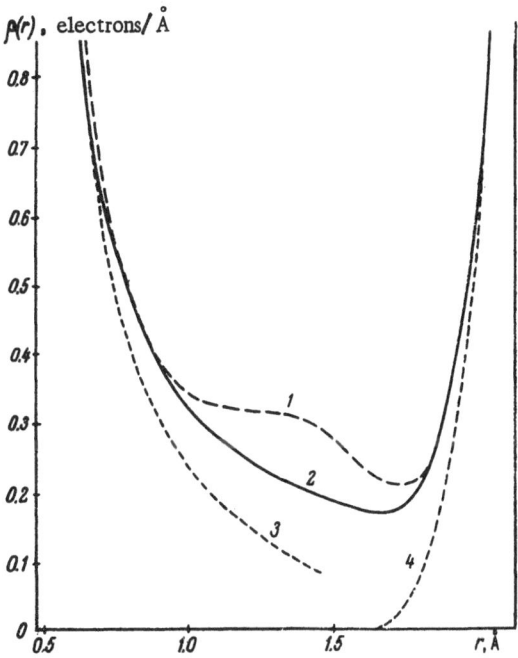

Fig. 3. Electron density in the [111] direction on the line joining an Si ion to its closest Mg neighbor, as calculated from the experimental structure factors F_{exp} (1) and the theoretical structure factors F_{th} (2), together with the distribution in the [100] direction, moving away from the Si and Mg atoms.

The results may be summarized as follows:

On the shortest connecting line between a silicon and a magnesium ion there is an electron bridge, indicating a partially covalent bond. Whereas the charge distribution in the neighborhood of the magnesium ion is nearly spherically symmetrical, the outer electrons of the silicon ion are displaced in the direction of the nearest magnesium ions. Determination of the degree of ionization shows that the magnesium ions carry 1.5 positive unit charges and the silicon ions carry one negative charge each. The remaining two valence electrons of the Mg_2Si formula unit are distributed equally between the ions.

The ionic radii found in our study lie in between the calculated Pauling values for Mg^{2+} and Si^{4-} and the covalent radii and are thus consistent with the homopolar type of bond here deduced.

The errors in the structure factors lie between 1 and 2%, and they correspond to average errors of 0.04 electron/$Å^3$ in the electron densities.

Literature Cited

1. R. Brill, H. Grimm, C. Hermann, and Cl. Peters, Ann. Phys. (Leipzig), 34:393 (1939).
2. H. Witte and E. Wölfel, Z. Phys. Chem. (Frankfurt), 3:296 (1955).

3.   J. Krug, H. Witte, and E. Wölfel, Z. Phys. Chem. (Frankfurt), 4:36 (1955).

4.   A. Weiss, H. Witte, and E. Wölfel, Z. Phys. Chem. (Frankfurt), 10:98 (1957).

5.   H. Bensch, H. Witte, and E. Wölfel, Z. Phys. Chem. (Frankfurt), 4:65 (1955).

6.   S. Göttlicher, Diploma Thesis, Darmstadt.

7.   S. Göttlicher and W. Wölfel, Z. Elektrochem., 63:891 (1959).

8.   E. Zintl, Z. Angew. Chem., 52:1 (1939).

9.   H. Welker, Ergeb. Exakten Naturwiss., 29:276 (1956).

10. H. Krebs, Acta Crystallogr., 9:95 (1956).

11. G. Busch and U. Winkler, Helv. Phys. Acta, 26:395 (1953).

12. N. V. Ageev and L. N. Guseva, Izv. Akad. Nauk SSSR, 1:31 (1952).

13. B. Wagner, Dissertation, Darmstadt (1953).

14. D. Panke, Diploma Thesis, Darmstadt (1963).

EXPERIMENTAL DETERMINATION OF THE STRUCTURE
AMPLITUDES AND OF THE DEGREE OF IONIZATION
IN GALLIUM PHOSPHIDE*

N. N. Sirota, A. U. Sheleg, and E. M. Gololobov

An x-ray diffraction determination was made of the atomic scattering factors of gallium and phosphorus ions in GaP. These functions were used to calculate the Debye–Waller factors B, the characteristic temperatures Θ, the mean-square dynamic displacements U_{dyn}^2, and the effective charges. The experimental values of the atomic scattering factors f were used to calculate the distribution of the electron density along the [111] direction in the lattice of GaP.

The x-ray diffraction method was used to determine the squares of the structure amplitudes of gallium phosphide, the atomic scattering factors of the gallium and phosphorus ions in this compound, and the distribution of the electron density in the lattice along the [111] direction.

The x-ray diffraction patterns were obtained using the URS-50IM diffractometer and the monochromatic Cu-K$_\alpha$ radiation. The monochromator consisted of a bent germanium plate, which was d = 0.17 mm thick. The plane of the plate was oriented along (111) and its radius of curvature was R = 330 mm. The monochromator was attached rigidly to the casing of the x-ray tube.

Adjusting screws were employed to rotate or displace the monochromator plate in mutually perpendicular planes.

A scintillation counter of the NaI:Tl type was used as the detector. A differential amplitude analyzer was employed as the discriminator. The x-ray diffraction spectrum was recorded either continuously or under fixed conditions. The integrated intensity of the reflections was calculated from the number of quanta recorded by an electromechanical counter.

The investigation was carried out on a fine powder of gallium phosphide whose grain size did not exceed 3μ. This powder was carefully ground in a jasper mortar and was then elutriated in toluene. The elutriated powder was used to prepare pellets 20 mm in diameter and 4–5 mm thick. The influence of the porosity and of the preferential orientation of the powder on the intensity of the x-ray lines was investigated by compacting the pellets at various pressures ranging from 6 to 8210 kgf/cm^2. The effect of porosity could be eliminated by increasing the compacting pressure. However, high pressures produced preferential orientation along some crystallographic planes.

---

* "Crystals," pp. 136–140 (see page 3).

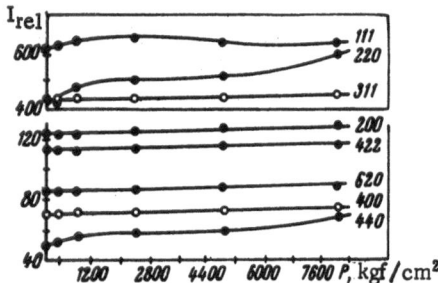

Fig. 1. Dependence of the relative integrated intensity I_{rel} of some reflections on the compacting pressure P.

It is evident from Fig. 1 that the relative integrated intensity of the majority of the investigated reflections (with the exception of 111, 220, and 440) was independent of the compacting pressure within the limits of the experimental error (1—2%). The intensity of the 111 line depended weakly on the pressure, whereas the intensities of the 220 and 440 reflections increased appreciably with the pressure. The increase in the 220 and 440 reflections was due to preferential orientation along the (110) planes because the relative intensity of the 440 line increased with the pressure proportionally to the intensity of the 220 line.

The structure amplitudes were measured using samples compacted by a pressure of 2460 kgf/cm². The relative intensities of the 220 and 440 reflections were corrected for the effect of preferential orientation. The intensities of the 400, 422, and 620 reflections were independent of the pressure and the square of the structure amplitudes for these reflections fitted a monotonic curve (Fig. 2). A comparison of the intensity of the 440 line found experimentally with the value of the intensity of the same line deduced from the F_1^2 curve (Fig. 2) was used to determine the correction necessary for the preferential orientation of the (110) planes. The same correction was applied to the intensity of the 220 reflection.

No correction was made for the extinction because the powder grains were sufficiently small. The influence of extinction was investigated earlier using Ge and InSb powders [8]. It was found that the relative intensity of the 111 reflection of Ge and InSb was practically independent of the degree of dispersion for particles whose size did not exceed 2-3 μ.

The polarization factor was calculated from the formula for a mosaic crystal:

$$P_f = (1 + \cos^2 2\vartheta_m \cdot \cos^2 2\vartheta)/(1 + \cos^2 2\vartheta_m).$$

The correction for the thermal diffuse scattering was made using the method of Chipman and Paskin [1].

The measured intensities of the diffraction lines were used to calculate the squares of the structure amplitudes of gallium phosphide (Fig. 2).

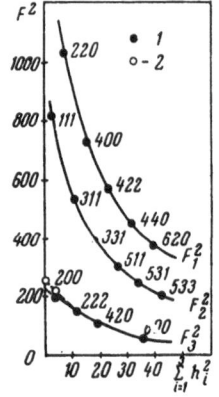

Fig. 2. Dependences of the squares of the structure amplitudes F^2 of gallium phosphide, reduced to absolute zero, on the sums of the squares of the indices $\sum_{i=1}^{3} h_i^2$: 1) experimental points; 2) theoretical values.

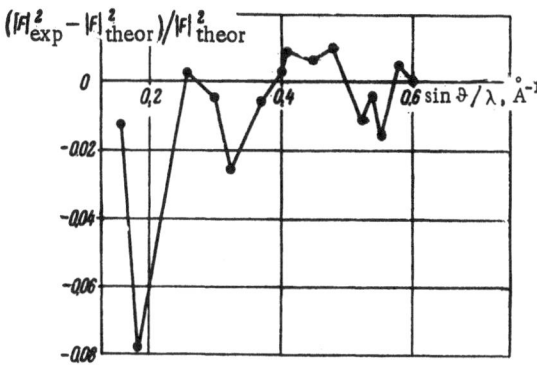

Fig. 3. Dependence of the relative difference between the experimental and the theoretical values of $|F|^2$ on $(\sin\vartheta)/\lambda$.

Figure 3 shows the relative difference between the experimental (reduced to absolute zero) and the theoretical values of $|F|^2$ as a function of the reflection order.

The values of the structure amplitudes were used to determine the atomic scattering factors f of the gallium and phosphorus ions. The absolute values of f were found by comparison with a standard. The standard was a fine-grained powder of nickel. The comparison line was 311. The Cromer dispersion correction [2] and the Debye–Waller temperature correction were applied to the functions f.

The experimental values of the atomic scattering factors (Fig. 4) and the tabulated trigonometric functions [3] were used, employing a method described in [4], to calculate the distributions of the electron density in the lattice of gallium phosphide along the [111] direction. It is evident from Fig. 5 that, as in the case of C, Si, and Ge [5], the electron density in the nearest

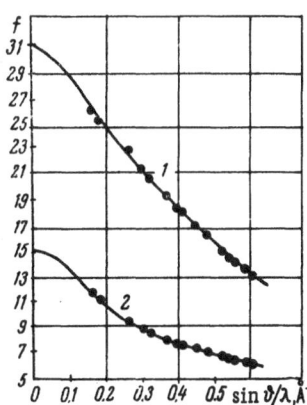

Fig. 4. Dependences, on $(\sin\vartheta)/\lambda$, of the experimental values (points) and the theoretical Hartree–Fock atomic scattering factors (continuous lines) of the gallium (1) and phosphorus (2) ions at T = 0°K.

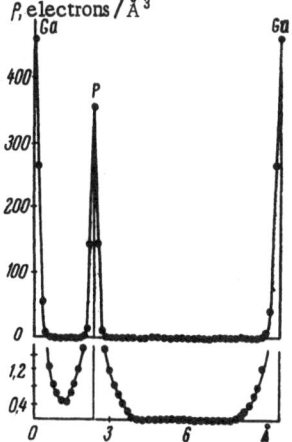

Fig. 5. Distribution of the electron density ρ along the [111] direction in the lattice of gallium phosphide.

Ga and P atoms increased to 0.43 electron/$\mathring{A}^3$. A comparison of the experimental atomic scattering factors of gallium and phosphorus with the theoretical Hartree–Fock factors [6] yielded the Debye–Waller factors B and the characteristic temperatures Θ. Their values were used to determine also the mean-square dynamic displacements $\overline{U^2_{dyn}}$. The following values were obtained for the gallium and phosphorus ions: Θ_{Ga} = 235°K, Θ_P = 360°K; B_{Ga} = 0.91 $\mathring{A}^2$, B_P = 0.89$\mathring{A}^2$; $(\overline{U^2_{dyn}})_{Ga}$ = 1.15 · 10^{-2} $\mathring{A}^2$, $(\overline{U^2_{dyn}})_P$ = 1.12 · 10^{-2} $\mathring{A}^2$.

The square of the structure amplitude for the 200 line, calculated per single molecule, was

$$|F|^2_{200} = (f_{Ga} - f_P)^2. \tag{1}$$

The experimentally determined value of the square of the structure amplitude for this line, reduced to absolute zero, was F^{2exp}_{200} = 203.9 ± 1, the theoretical value calculated from the f curves of gallium and phosphorus [6] was F^{2theor}_{200} = 220.5, and the square of the structure amplitude for these reflections was F^{2theor}_{000} = 256 for $(\sin\vartheta)/\lambda$ = 0.

The open circles of Fig. 2 represent the theoretical values of F^2_3. The experimental curve representing F^{2exp}_3 (black dots) was extrapolated to $(\sin\vartheta)/\lambda$ = 0. It is evident from Figs. 2 and 3 that at small values of $(\sin\vartheta)/\lambda$ the difference between the experimental and the theoretical F^2_3 curves exceeded considerably the experimental error (1–2%) in the determination of the F^{2exp}_3 curve. This difference amounted to 8% for the 200 line and $(\sin\vartheta)/\lambda$ = 0.183. At low reflection angles, the experimental curve F^{2exp}_3 lay well below the theoretical curve. It was suggested in [7, 9] that this was evidence of the transfer of electrons from the atoms of the metal to the atoms of the metalloid. We were unable to determine experimentally the value of F^2_3 for $(\sin\vartheta)/\lambda$ = 0 and, therefore, this value was deduced from the intersection of the extrapolated F^{2exp}_3 curve with the ordinate: $F^{2\,exp}_{3\,000}$ = 225 ± 4. The degree of ionization in GaP was deduced by a method described earlier [7], using the following formula

$$[(f^{theor}_{Ga} - x) - (f^{theor}_P + x)]^2_{000} = F^{2\,exp}_{3\,000}, \tag{2}$$

where x is the degree of ionization expressed in electrons; f^{theor}_{Ga} and f^{theor}_P are the theoretical values of the atomic scattering factors of gallium and phosphorus for $(\sin\vartheta)/\lambda$ = 0, which could be replaced by the atomic numbers. Therefore,

$$Z_{Ga} - x - Z_P - x = F^{exp}_{3\,000},$$
$$x = \frac{Z_{Ga} - Z_P - F^{exp}_{3\,000}}{2}. \tag{3}$$

The degree of ionization (the effective charge of the ions) in gallium phosphide calculated from Eq. (3) was x = 0.5 ± 0.10 electron. The presence of an electron "bridge" and the effective charges of the ions indicated that the bonding in gallium phosphide was of the mixed ionic–covalent type.

Literature Cited

1.   D. R. Chipman and A. Paskin, J. Appl. Phys., 30:1998 (1959).
2.   D. T. Cromer, Acta Crystallogr., 18:17 (1965).
3.   I. M. Kuntsevich, N. M. Olekhnovich, and A. U. Sheleg, Tables of Trigonometric Functions for Calculations of the Electron Density Distribution in Crystals [in Russian], Nauka i Tekhnika, Minsk (1967).
4.   N. N. Sirota, N. M. Olekhnovich, and A. U. Sheleg, Dokl. Akad. Nauk SSSR, 132:160 (1960).
5.   A. U. Sheleg, Vestsi Akad. Navuk Belarus. SSR, Ser. Fiz.-Tekh. Navuk, No. 2, p. 51 (1964).

6.   International Tables for X-Ray Crystallography, Vol. 3, Kynoch Press, Birmingham, England (1962).

7.   N. N. Sirota, E. M. Gololobov, N. M. Olekhnovich (Olechnovich), and A. U. Sheleg, Krist. Tech., 1:545 (1966).

8.   N. N. Sirota, Acta Crystallogr., A25:223 (1969).

9.   N. N. Sirota and E. I. Gololobov, Dokl. Akad. Nauk SSSR, 156:1071 (1964).

MEAN-SQUARE DYNAMIC DISPLACEMENTS AND ATOMIC SCATTERING FACTORS OF IONS IN ALUMINUM NITRIDE*

N. N. Sirota, A. I. Olekhnovich, and N. M. Olekhnovich

The absolute values of the squares of the structure amplitudes (F^2) were determined for AlN in the temperature range 85-670°K using monochromatic Cu K_α radiation. These values were used to calculate the mean-square dynamic displacements and the atomic scattering factors of the Al and the N ions. The values of F^2 were used also to find the shortest relative distance (u_0/c) between the Al and the N ions along the c axis. This distance was 0.386 ± 0.001, which is different from 0.375 for a perfect structure ($c/a = 1.633$) and from 0.380 for the case of equal values of all the shortest atomic spacings ($c/a = 1.600$). The temperature dependences indicated that the mean-square dynamic displacements ($\overline{u^2}$) in AlN were anisotropic. Thus, at room temperature, these displacements were $\overline{u_{xy}^2} = (0.30 \pm 0.02) \cdot 10^{-2} \, \text{Å}^2$, $\overline{u_z^2} = (0.65 \pm 0.03) \cdot 10^{-2} \, \text{Å}^2$ for the Al ion and $\overline{u_{xy}^2} = (0.52 \pm 0.02) \cdot 10^{-2} \, \text{Å}^2$, $\overline{u_z^2} = (1.00 \pm 0.03) \cdot 10^{-2} \, \text{Å}^2$ for the N ion. The linear expansion coefficient (α) was also anisotropic. In the temperature range 290-670°K, the components of this coefficient were $\alpha_z = (3.0 \pm 0.2) \cdot 10^{-6} \, \text{deg}^{-1}$ and $\alpha_{xy} = (3.8 \pm 0.2) \cdot 10^{-6} \, \text{deg}^{-1}$. The experimental values F_{exp}^2 were extrapolated to absolute zero and compared with the theoretical values F_{theor}^2. It was found that F_{exp}^2 was less than the theoretical value due to the partial redistribution of the valence electrons away from the aluminum toward the nitrogen.

Aluminum nitride (AlN) differs from other $A^{III}B^V$ compounds in its hexagonal wurtzite structure. The lattice parameters of this compound are $a = 3.111$ Å, $c = 4.978$ Å, and $c/a = 1.600$ [1]. Thus, the structure of aluminum nitride differs from the perfect wurtzite structure, which consists of regular tetrahedra and is characterized by the axial ratios $c/a = 1.633$. Moreover, aluminum nitride has distinctive physical and physicochemical properties.

These properties make it desirable to study the nature of the chemical bonding and the crystal lattice dynamics of this compound.

The purpose of our investigation was to determine the mean-square dynamic displacements and the atomic scattering factors of the aluminum and the nitrogen ions in aluminum nitride. This was done by measuring the intensities of the x-ray diffraction spectra in the temperature range 85-670°K.

Aluminum nitride was synthesized from ammonia using the gas-transport reaction method. This yielded single-crystal "whiskers" about 1-2 mm long and a few tenths of a millimeter in diameter. The x-ray diffraction study was carried out on transparent single crystals (whiskers)

---

* "Crystals," pp. 141-149 (see page 3).

TABLE 1. Reduced Values of the Squares of the Structure Amplitudes of Aluminum Nitride

hkl	F^2 at 19°C	$\tilde{F}^2$ at 19°C	n	$\tilde{F}^2$ 0°K	$\tilde{F}^2_{theor}$	$\dfrac{\lvert \tilde{F}^2_{exp} - \tilde{F}^2_{theor} \rvert \cdot 100}{\tilde{F}^2_{exp}}$
010	199.0±1.0	199.0±1.0	1	203.0±1.0	204.77	0.87
002	449.2±2.8	112.3±0.7	4	115.3±0.7	115.96	0.58
011	113.1±1.5	37.7±0.5	3	37.8±0.5	40.55	7.30
012	78.2±0.8	78.2±0.8	1	85.1±0.9	87.40	2.70
110	411.6±4.4	102.9±1.1	4	111.7±1.1	113.85	1.93
013	210.0±2.1	70.0±0.7	3	80.9±0.7	82.61	2.11
020	84.3±2.0	84.3±2.0	1	92.9±2.0	94.09	1.28
112	210.0±0.8	52.5±0.2	4	58.9±0.2	60.25	2.28
021	76.5±0.6	25.5±0.2	3	27.8±0.2	28.06	0.93
004	74.0±7.6	18.5±1 9	4	21.3±2.0	20.30	4.58
022	43.2±1.0	43.2±1.0	1	48.6±1.0	50.11	3.11
014	17.0±1.3	17.0±1.3	1	19.6±1.3	17.94	8 15
023	128.4±1.5	42.8±0.5	3	51.6±0.5	52.71	2.14
120	51.8±0.7	51.8±0.7	1	60.3±0.7	60.84	0.90
121	50 1±0.9	16.7±0.3	3	18.6±0.3	18.21	2.20
114	46.8±1.6	11.7±0.4	4	13.7±0 4	13.70	0.0
122	29.2±0.7	29.2±0.7	1	35.3±0.7	35.68	1.08
015	108.6±2.1	36.2±0.7	3	50.6±0.7	50.25	0.69
024	9.5±0.4	9.5±0.4	1	11.4±0.4	12.65	11.0
030	156.4±3.2	39.1±0.8	4	48.5±0.8	48.02	0.99
123	86.4±1.8	28.8±0.6	3	38.6±0.6	37.28	3.41
032	80.4±1.6	20.1±0.4	4	25.4±0 4	28.34	11.5
006	44.8±5.6	11.2±1.4	4	16.4±1.5	18.32	11.7

which had been ground in a jasper mortar. The fine powder obtained in this way was elutriated in toluene so as to select the fraction having a grain size of 2-3 μ. The finer fraction of the powder was used to prepare compacted pellets, which were employed in investigations of the influence of temperature on the x-ray diffraction patterns. The transparency of the crystals used in the x-ray diffraction studies indicated that their composition was close to stoichiometric. The flat pellets did not exhibit a noticeable texture.

At low and high temperatures, the x-ray diffraction measurements were carried out in a vacuum chamber.

The intensities of the x-ray diffraction peaks were determined using a diffractometer with a monochromatic Cu K_α source. The intensities were measured with an NaI:Tl scintillation counter coupled to a discriminator. The copper radiation was made monochromatic by an elastically bent single-crystal germanium plate cut along the (111) plane [2].

The intensities of the reflections were deduced from the number of pulses recorded at points with an angular separation equal to the width of the counter slit (43'). In this method, the number of pulses counted in a given (short) time interval was higher than that in continuous recording and this improved the accuracy of the measurements of the integrated intensity of the reflections.

The influence of temperature was investigated in two stages. We measured the intensities of the diffraction peaks at room temperature without recourse to an x-ray vacuum camera. The integrated intensities measured at 19°C were used to calculate the squares of the structure amplitudes $\lvert F \rvert^2$ (Table 1). The values of $\lvert F \rvert^2$ were converted to absolute units by measuring the intensity of the primary peak (I_0) and comparing it with the intensities of the reflections. Standard attenuation filters were used in the measurements of the primary peak intensity.

Next, the sample was inserted into a vacuum diffractometer camera which was placed on a goniometer. All the diffraction lines were measured at room, low, and high temperatures. The ratios of the reflection intensities, measured at some temperature T, to the intensities of

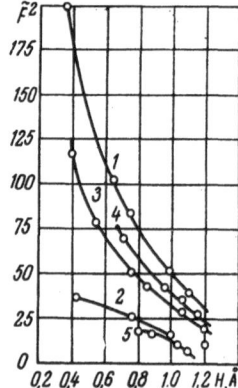

Fig. 1. Dependences of the reduced values of the squares of the structure amplitude $\widetilde{F}^2$ determined for aluminum nitride at 19°C, on $H = 2(\sin\vartheta)/\lambda$, for various values of l: 1) 0; 2) 1; 3) 2; 4) 3; 5) 4.

the same reflections measured at room temperature were determined and used to calculate the squares of the structure amplitudes. The angular intensity factors were corrected to the thermal shifts of the diffraction lines.

Such calculations showed that the square of the structure amplitude $|F|^2$ for aluminum nitride powder was given by the following relationships:

$$|F|^2 = \frac{2}{3}\left[f_{Al}^2 + f_N^2 + 2f_{Al}f_N \cos 2\pi \frac{u_0}{c} l \right] \times$$

$$\times \left\{ 3 + (-1)^l \left[4\cos \frac{2\pi}{3} \frac{h-k}{2} \cos \frac{2\pi}{3} \frac{2h+k}{2} \cos \frac{2\pi}{3} \frac{h+2k}{2} - 1 \right] \right\}. \tag{1}$$

This relationship could be represented also in the form

$$|F|^2 = n|\tilde{F}|^2, \tag{1a}$$

where

$$n = \frac{2}{3}\left\{ 3 + (-1)^l \left[4\cos \frac{2\pi}{3} \frac{h-k}{2} \cos \frac{2\pi}{3} \frac{2h+k}{2}, \cos \frac{2\pi}{3} \frac{h+2k}{2} - 1 \right] \right\}$$

is a quantity which is determined by the value of the Miller indices and is independent of the atomic scattering factors of the ions in the lattice; $|\widetilde{F}|^2$ is the reduced value of the square of the structure amplitude,

$$|\tilde{F}|^2 = \left[f_{Al}^2 + f_N^2 + 2f_{Al}f_N \cos 2\pi \frac{u_0}{c} l \right], \tag{2}$$

which depends on the atomic scattering factors f_{Al} and f_N, on the parameter u_0/c (the relative distance between the Al and the N ions along the c axis); and on the Miller index l of the reflecting plane (hk$l$).

It is evident from Eq. (2) that the reduced values of the square of the structure amplitude $|\widetilde{F}|^2$ can be divided into groups characterized by $l = 0, 1, 2, 3, 4$, etc. (Fig. 1). Moreover, an examination of Eq. (2) shows that the dependence of $|\widetilde{F}|^2$ on u_0/c is not the same for different values of l. For example, when u_0/c is increased, the intensities of the reflection characterized by $l = 1$ and $l = 3$ decrease, whereas the intensity of the $l = 2$ reflection increases. An analysis of this type has established that the quantity

$$g = \frac{|\tilde{F}|_{l=0}^2 - |\tilde{F}|_{l=1}^2}{|\tilde{F}|_{l=3}^2 - |\tilde{F}|_{l=2}^2} = \frac{1 - \cos 2\pi u_0/c}{\cos 2\pi \frac{3u_0}{c} - \cos 2\pi \frac{2u_0}{c}} \tag{3}$$

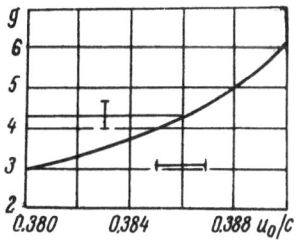

Fig. 2. Dependence of the parameter g on the relative distance u_0/c between the Al and the N ions along the c axis.

is quite sensitive to the parameter u_0/c (Fig. 2) and can be used in the experimental determination of this parameter. It is sufficient to calculate the values of $|\widetilde{F}|^2$ from the monotonic curves of Fig. 1 and then compute the value of g. The parameter u_0/c is then found directly from the curve plotted in Fig. 2. The error in the determination of u_0/c resulting from the anisotropy of the temperature factor can be reduced by selecting the values of $|\widetilde{F}|^2$ in the range of low and moderate values of the reflection order.

This method was used to calculate the nearest distance (u_0/c) between the Al and the N ions along the c axis from the measured values of the square of the structure amplitude. We found (Fig. 2) that $u_0/c = 0.386 \pm 0.001$ and that it was independent of temperature (within the limits of the experimental error). Jeffrey et al. [1] reported 0.385 for this parameter. If all the shortest atomic spacings were equal and $c/a = 1.600$, we should obtain $u_0/c = 0.380$, whereas in the case of a perfect wurtzite structure ($c/a = 1.633$) we should have 0.375. Hence, we concluded that the ions located within tetrahedra in the aluminum nitride lattice were displaced somewhat toward the base of the unit cell. The shortest distance between the Al and the N ions, calculated for $a = 3.111$ Å and $c = 4.978$ Å, was 1.922 Å along the c axis and 1.883 Å along the three other directions.

We measured the unit-cell parameters (a and c) of aluminum nitride and found that they agreed, within 0.001 Å, with those given by Jeffrey et al. [1].

Measurements of the diffraction peak intensities at various temperatures indicated that the vibrations of the ions in the AlN lattice were anisotropic. The intensities of different diffraction lines depended differently on the temperature (Fig. 3). Variation of the temperature affected most the lines with large values of the Miller index l and least the lines characterized by $l = 0$. This indicated that the vibrations of the ions had a larger amplitude along the c axis than in the basal plane. Under these conditions, the atomic scattering factors of the Al and the N ions should also be anisotropic, i.e.,

$$f_{jxy} = f_{0j}\, e^{-2\pi^2 \overline{u^2_{jxy}} H^2_{xy}},$$
$$f_{jz} = f_{0j}\, e^{-2\pi^2 \overline{u^2_{jz}} H^2_{z}},$$

(4)

so that

$$f_{jT} = f_{0j}\, e^{-2\pi^2 \overline{u^2_{jxy}} H^2_{xy} - 2\pi^2 \overline{u^2_{jz}} H^2_{z}},$$

(5)

Fig. 3. Temperature dependences of $\ln|\widetilde{F}|^2$ of some diffraction lines observed at large reflection angles: 1) 300 ($H^2 = 1.24$); 2) 302 (1.40); 3) 213 (1.33); 4) 114 (1.06); 5) 204 (1.20); 6) 105 (1.15).

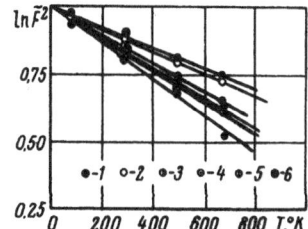

where f_{0j} is the atomic scattering factor of the j-th ion at rest; $\overline{u^2_{jxy}}$ and $\overline{u^2_{jz}}$ are the mean-square dynamic displacements of the j-th ion in the basal plane of the unit cell and along the z axis, respectively; $H^2_{xy} = 4/3a^2(h^2 + hk + k^2)$; $H^2_z = (l^2/c^2)$. When Eqs. (4) and (5) are taken into account, the expression for the reduced square of the structure amplitude (2) becomes:

$$|\bar{F}|^2 = f^2_{0Al}\, e^{-4\pi^2 \overline{u^2_{Alxy}} H^2_{xy} - 4\pi^2 \overline{u^2_{Alz}} H^2_z} + f^2_{0N}\, e^{-4\pi^2 \overline{u^2_{Nxy}} H^2_{xy} - 4\pi^2 \overline{u^2_{Nz}} H^2_z} + 2k f_{0Al} f_{0N}\, e^{-2\pi^2 (\overline{u^2_{Alxy}} + \overline{u^2_{Nxy}}) H^2_{xy} - 2\pi^2 (\overline{u^2_{Alz}} + \overline{u^2_{Nz}}) H^2_z}, \quad (6)$$

where $k = \cos 2\pi\, (u_0/c)l$.

It is evident from Eq. (6) that, for the $l = 0$ and $l = 1$ curves (Fig. 1) and for the reflection orders which satisfy $H^2_{z(l=1)} \ll H^2_{xy(l=1)}$, the reduced values of the square of the structure amplitude are practically independent of the mean-square dynamic displacements of the ions along the z axis:

$$\left.\begin{array}{l} |\bar{F}|^2_{l=0} = f^2_{Alxy} + 2f_{Alxy} f_{Nxy} + f^2_{Nxy} \\[4pt] |\bar{F}|^2_{l=1} \cong f^2_{Alxy} + 2k f_{Alxy} f_{Nxy} + f^2_{Nxy} \quad (\text{if } H^2_z \ll H^2_{xy}) \end{array}\right\}. \quad (7)$$

The solution of the system (7) gives the values of f_{Alxy} and f_{Nxy} at a given temperature. The values of f_{Alxy} and f_{Nxy} determined in this way for different temperatures can be used to calculate the mean-square dynamic displacements $(\overline{u^2_{xy}})$ of the aluminum and the nitrogen ions in the basal plane of the unit cell (these displacements can be calculated separately for aluminum and nitrogen).

Using the values of $\overline{u^2_{xy}}$, calculated in this way, we can easily determine the mean-square dynamic displacements of the Al and the N ions along the z axis from the values of $|\tilde{F}|^2$ for $l = 2$ and 3.

The results of such calculations of $\overline{u^2_{xy}}$ and $\overline{u^2_z}$ show that the mean-square dynamic displacements of the aluminum and the nitrogen ions along the z axis are approximately twice as large as in the basal plane (Fig. 4). The room-temperature displacements of the aluminum ions are $\overline{u^2_{xy}} = (0.30 \pm 0.02) \cdot 10^{-2}\ \text{Å}^2$, $\overline{u^2_z} = (0.65 \pm 0.03) \cdot 10^{-2}\ \text{Å}^2$, and those of the nitrogen ions are $(0.52 \pm 0.02) \cdot 10^{-2}\ \text{Å}^2$ and $(1.00 \pm 0.03) \cdot 10^{-2}\ \text{Å}^2$, respectively. In the Debye theory approximation, these values correspond to the following characteristic temperatures: $\Theta_{xy} = 790$, $\Theta_z = 510°K$ for Al and $\Theta_{xy} = 850$, $\Theta_z = 580°K$ for N.

It is evident from Fig. 5 that the ratios $\overline{u^2_z}/\overline{u^2_{xy}}$ for the aluminum and the nitrogen ions increased with rising temperature. Measurements of the linear expansion coefficient (α) by the x-ray diffraction method showed that this coefficient was also anisotropic. In the temperature range 298–670°K, the linear expansion coefficient along the c axis (α_z) was $(3.0 \pm 0.2) \cdot 10^{-6}$ deg^{-1}, and in the basal plane it was $\alpha_{xy} = (3.8 \pm 0.2) \cdot 10^{-6}$ deg^{-1}. The ellipsoid of the square of the linear expansion coefficient was elongated in the basal plane (Fig. 4).

The calculated values of the mean-square dynamic displacements of the ions in the AlN lattice were used to find the squares of the structure amplitudes at absolute zero, i.e., the measured values of $|\tilde{F}|^2$ were corrected for the influence of the temperature (Table 1). Since the temperature factor was anisotropic, the experimental values of $|\tilde{F}|^2$ were corrected for the temperature using the following relationship:

$$|\bar{F}|^2_{0\,exp} = |\bar{F}|^2_{T\,exp} \frac{|F|^2_{0\,theor}}{|\bar{F}|^2_{T\,theor}}, \quad (8)$$

where $|\tilde{F}|^2_{T\,exp}$ and $|\tilde{F}|^2_{0\,exp}$ are the experimental values of the reduced square of the structure amplitude at a temperature T and at absolute zero (the second value takes into account the zero-point vibrations); $|\tilde{F}|^2_{T\,theor}$ and $|\tilde{F}|^2_{0\,theor}$ are the corresponding theoretical values of the reduced square of the structure amplitude at T and 0°K.

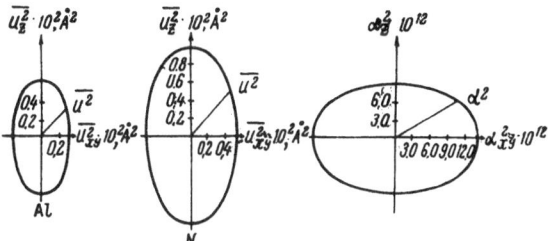

Fig. 4. Ellipsoids of mean-square dynamic displacements of the aluminum and the nitrogen ions at 19°C and the ellipsoid of the squares or components of the linear expansion coefficient.

The values of $|\tilde{F}|^2_{T\,\text{theor}}$ and $|\tilde{F}|^2_{0\,\text{theor}}$ were calculated using the theoretical values of the Hartree–Fock f curves for neutral atoms of aluminum and nitrogen [4]. The values of $|\tilde{F}|^2_{T\,\text{theor}}$ were calculated at room temperature employing the reduced values of the mean-square dynamic displacement.

The experimental values obtained in this way are listed in Table 1. It is evident from Fig. 6 that the experimental and the theoretical values of $|\tilde{F}|^2$ are in good agreement. The agreement between $|\tilde{F}|^2_{\text{exp}}$ and $|\tilde{F}|^2_{\text{theor}}$ at high reflection orders indicate the absence of significant static displacements of the ions. However, it must be mentioned that the experimental value of $|\tilde{F}|^2$ for the (011) reflection is less than the theoretical value. The difference between the experimental and the theoretical values of $|\tilde{F}|^2$ for the (011) reflection exceeds the experimental error by a factor of about 5.6. The discrepancies for the other reflections are much smaller and do not exceed twice the value of the experimental error.

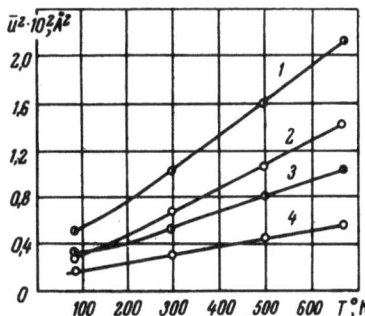

Fig. 5. Temperature dependence of the mean-square dynamic displacements of the aluminum and the nitrogen ions in the AlN lattice: 1) $\overline{u^2_{zN}}$; 2) $\overline{u^2_{zAl}}$; 3) $\overline{u^2_{xyN}}$; 4) $\overline{u^2_{xyAl}}$.

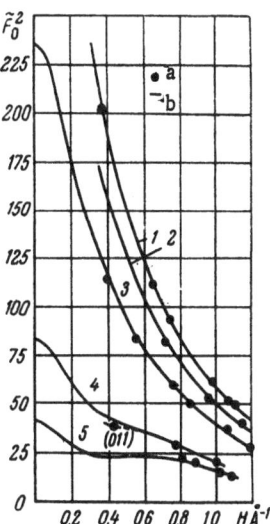

Fig. 6. Dependences of the reduced values of the square of the structure amplitude $\tilde{F}^2_0$ of aluminum nitride at 0 °K on H = 2(sinϑ)/λ for various values of l: 1) 0; 2) 3; 3) 2; 4) 1; 5) 4. a) Experimental values; b) theoretical values.

The value of $|\widetilde{F}|^2$ for the (011) reflection is, in practice, determined by the difference between the atomic scattering factors of the aluminum and the nitrogen ions and the reflection itself is observed at relatively low reflection orders. This means that the (011) reflection is sensitive to the state of the ions in the crystal lattice. The observed decrease in $|\widetilde{F}|^2$ for the (011) reflection is evidently due to a partial redistribution of the valence electrons resulting in their shift from aluminum to nitrogen [5].

The values obtained for the dynamic displacements of the ions and of the structure factors may be used in considering the problem of atomic bonding in aluminum nitride.

Literature Cited

1. G. A. Jeffrey, G. S. Parry, and R. L. Mozzi, J. Chem. Phys., 25:1024 (1956).
2. N. N. Sirota and N. M. Olekhnovich, Vestsi Akad. Navuk Belarus. SSR, Ser. Fiz.-Tekh. Navuk, No. 1, p. 111 (1968).
3. R. W. James, The Optical Principles of the Diffraction of X-Rays, Bell, London (1950).
4. International Tables for X-Ray Crystallography, Vol. 3, Kynoch Press, Birmingham (1962).
5. N. N. Sirota, E. M. Gololobov, and N. M. Olekhnovich (Olechnovič), and A. U. Sheleg (Šeleg), Krist. Tech., 1:545 (1966).

X-RAY SPECTROSCOPIC INVESTIGATION OF CHEMICAL BONDING IN SOME RARE-EARTH TITANATES*

A. Meisel and G. Leonhardt

Leipzig, Germany

An investigation was made of the position, width, and asymmetry of the $K\alpha_{1,2}$ lines of titanium in various chemical compounds. The measurements were made in a focusing vacuum x-ray spectrometer using the fluorescent excitation method. The results obtained made it possible to analyze the valence of titanium in some rare-earth titanates. Use was made of the shift of the $K\alpha_1$ titanium lines, as well as of their width and asymmetry for different valence states (Ti^{2+}, Ti^{3+}, Ti^{4+}). A comparison was made of the x-ray spectroscopic, the magnetic, and the structure data for the compounds investigated. The advantages of the x-ray spectroscopic method or the magnetic method were demonstrated in the determination of the valence in the specific case of $EuTi_2O_4$.

We recently carried out a systematic investigation of the influence of chemical bonding on the positions and the profiles of the $K\alpha_{1,2}$ emission lines of the 3d transition elements. All these elements exhibited a definite dependence of the x-ray wavelength on their valence and on the electronegativity of the neighboring atoms. The width and the asymmetry of the lines depended on the number of unpaired electrons and this was also true of the magnetic moments of simple compounds. These results enabled us to obtain much information on chemical bonding, particularly in the case of complex compounds.

In the case of titanium, our aim was to determine the valence of some rare-earth titanates by the x-ray spectroscopic method. The secondary-reflection $K\alpha_{1,2}$ titanium lines were recorded photographically by means of a high-vacuum x-ray spectrograph with Johann focusing. An x-ray beam was reflected from the $(10\bar{1}1)$ plane of quartz or, in some cases, from a cleavage plane of mica (fifth-order reflection). The radii of curvature of the bent crystals were 1002 and 1217 mm. The dispersion achieved using these crystals was 2.33 XU/mm (i.e., 4.01 eV/mm) and 3.59 XU/mm (6.01 eV/mm). This enabled us to resolve completely the $K\alpha_{1,2}$ doublet whose components were separated by 3.65 XU. A crystal was bent and the aperture was reduced until we achieved the lowest value of the half-width of the $K\alpha_1$ line of metallic titanium ($\Delta\lambda = 0.93$ XU). The asymmetry index of this line was $\alpha = 1.13$.

The titanium compounds were excited by the secondary method, in which the maximum

---

* "Crystals," pp. 150–154 (see page 3).

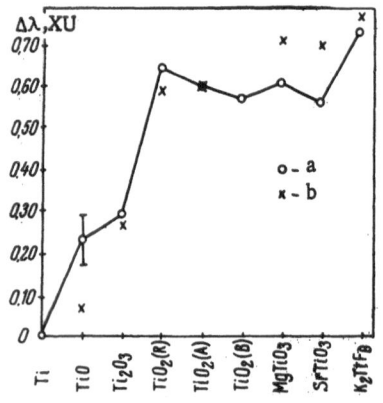

Fig. 1. Shifts of the K_{α_1} line of
Ti: a) [1]; b) [2].

power supplied to the x-ray tube was 0.6 kW
(30 kV and 20 mA). An exposure of 1 h for the
titanium and of 3–13 h for the compound was
required to obtain the optical densities lying in
the linear part of the characteristic of the photographic film. Three bands were recorded in
succession in each film: the $K_{\alpha_{1,2}}$ doublet of
the compound was recorded above and below the
central position occupied by the doublet of the
titanium metal. The $K_{\alpha_{1,2}}$ lines of metallic
copper (ninth reflection order) were used for
calibration purposes.

The valences were determined from the
shift of the K_{α_1} line of titanium, from the broadening index $\nu = (\Delta\lambda)_c / (\Delta\lambda)_e$ (the subscripts
c and e are used for the compounds and the element, respectively), and from the relative
change in the asymmetry index of this line $\alpha = (\alpha_c - \alpha_e)/\alpha_e$ [4]. Figure 1 shows the total
shifts of the lines. The values obtained by Blokhin and Shuvaev [2] are also included (the value
for $CaTiO_3$ was used instead of $MgTiO_3$). Considerable differences were observed in the positions of the lines of trivalent and tetravalent titanium (the experimental error was 0.06 XU).
Figures 2 and 3 show the broadening and the asymmetry indices which were used to separate
the trivalent and the divalent titanium states. The considerable broadening of the line of
K_2TiF_6 is due to the high electronegativity of fluorine.

The compounds $LaTiO_3$ and $EuTiO_3$ were prepared by heating a stoichiometric mixture
of Ti_2O_3 with La_2O_3 or Eu_2O_3 at 1150°C for 3 days. Moreover, $EuTiO_3$ was prepared by firing
a mixture of Eu_2O_3 and TiO_2, taken in the ratio 1:2, in a hydrogen atmosphere at 1200°C and by
firing a mixture of EuO and TiO_2 in an evacuated quartz ampoule. The EuO was prepared by
the reaction between fine powders of EuOCl and LiH in high vacuum at temperatures from 600
to 800°C; and the Ti_2O_3 was prepared by prolonged heating of equimolar amounts of TiO and
TiO_2 in a well-evacuated quartz ampoule at 1100°C. $LaTi_2O_4$ was prepared by firing $LaTiO_3$
and TiO_2 in a hydrogen atmosphere. $EuTi_2O_4$ and $SrTi_2O_4$ were obtained by heating a mixture
of Ti_2O_3 with EuO or SrO. Eu_2TiO_4 was prepared similarly from Eu_2O_3 and TiO. All the
titanates were homogeneous black powders [3].

Table 1 lists the results of our x-ray spectroscopic investigations. The first three rows
of the table give the average values of the shift and the changes in the profile of the K_{α_1} line of

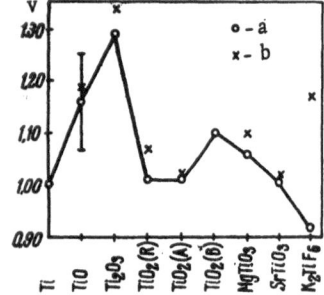

Fig. 2. Broadening indices
of the K_{α_1} line of Ti: a) [1];
b) [2].

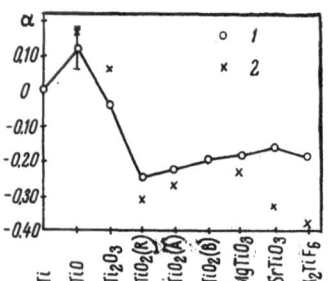

Fig. 3. Relative changes
in the asymmetry index
of the K_{α_1} line of Ti: 1)
[1]; 2) [2].

TABLE 1. X-Ray Spectroscopic, Magnetic, and Structure Data on Rare-Earth Titanates

Compound	$\delta\lambda$, XU	v	a	u_{Ti}	u_{tot}	μ eff, μ_B	Nominal valences	a	b	c	Structure
									Å		
1	2	3	4	5	6	7	8	9	10	11	12
Ti^{2+}	+0.23	1.16	+0 12	2	2						
Ti^{3+}	+0.29	1.29	−0.04	1	1	0.9— 1.2					
Ti^{4+}	+0.63	1.02	−0.20	0	0	0					
$LaTi_2O_4$		1.19	+0.01	1.5	3	0.8	$La^{III}Ti^{II}(Ti^{III}O_4)$	3.90	3.95	12.80	K_2NiF_4,
$EuTi_2O_4$		1.34	−0.04	1	9	8.3	$Eu^{II}Ti^{III}(Ti^{III}O_4)$	3.95	3.97	12.75	orthorhomb.
$SrTi_2O_4$		1.30	−0.08	1	2	1.9	$Sr^{II}Ti^{III}(Ti^{III}O_4)$	3.89	3.98	12.76	distorted
$LaTiO_3$	+0.39	1.20	−0.06	1	1	1.3	$La^{III}Ti^{III}O_3$	5.55	5.75	7.83	$CaTiO_3$, orthorhomb. distorted
$EuTiO_3$	+0.62	1.16	−0.14	0	7	7.7	$Eu^{II}Ti^{IV}O_3$	7.81			$CaTiO_3$
$MgTiO_3$	+0.61	1.06	−0.18	0	0	0	$Mg^{II}Ti^{IV}O_3$	5.54			$FeTiO_3$
$SrTiO_3$	+0.56	1.01	−0.16	0	0	0	$Sr^{II}Ti^{IV}O_3$	3.90			$CaTiO_3$
Eu_2TiO_4	+0.51	1.21	−0 13	0	14	8.9	$Eu_2^{II}Ti^{IV}O_4$	3.88	4.00	12.75	K_2NiF_4, orthorhomb. distorted

Ti in various valence states [1]. Columns 5 and 6 list the theoretical numbers of unpaired electrons u_{Ti} per one titanium atom and the nominal number of all the unpaired electrons u_{tot} in the compounds investigated. Column 7 gives the magnetic moment μ_{eff} in Bohr magnetons (μ_B) measured by the Gouy method [3]. The valences of the various titanates deduced from these investigations are given in column 8. The data obtained for $LaTi_2O_4$ can be explained on the basis of a band model, in which a nominal degree of ionization of 2.5 is attributed to both titanium ions. The titanium in $SrTi_2O_4$, $EuTi_2O_4$, and $LaTi_2O_4$ is present in the trivalent state.

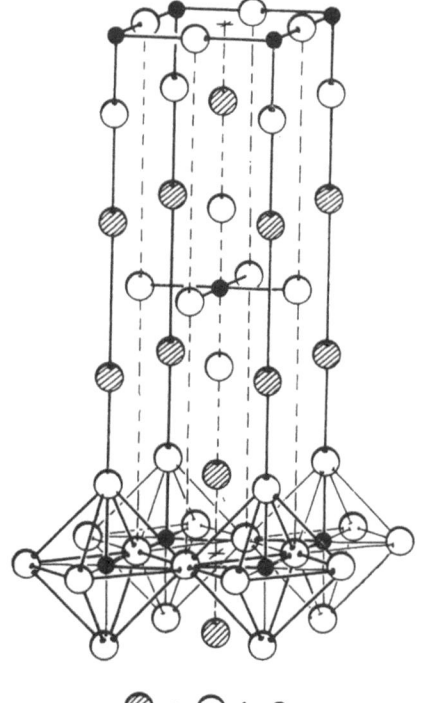

Fig. 4. Crystal structure of K_2NiF_4: a) potassium ions; b) fluorine ions; c) nickel ions.

a   b   c

In the case of $EuTi_2O_4$, we can see clearly the advantages of using the x-ray spectroscopic method in conjunction with the magnetic method for the determination of valence since all three possible distributions of the valence in this compound, (a) $Eu^{II}Ti^{III}Ti^{III}$, (b) $Eu^{II}Ti^{II}Ti^{IV}$, and (c) $Eu^{III}Ti^{II}Ti^{III}$, lead to nine unpaired electrons and, therefore, to practically the same magnetic moment. However, the width and the asymmetry of the K_{α_1} line of titanium give preference to the distribution (a). $EuTiO_3$, Eu_2TiO_4, $MgTiO_3$, and $SrTiO_3$ contain tetravalent titanium. A comparison of the K_{α_1} line of Ti in $EuTiO_3$ and $EuTi_2O_4$ with the corresponding line in $SrTiO_3$ and $SrTi_2O_4$ show clearly the influence of paramagnetic Eu on the width of the Ti line.

The valences of the titanium ions found by the x-ray spectroscopy method are in agreement with the x-ray diffraction data obtained by the Debye–Scherrer and Guinier methods, or by employing an x-ray interference goniometer [3]. The reflections obtained may be attributed to an orthorhombic lattice when the extinction rule is taken into account. The Miller indices give the lattice constants (a, b, c) listed in columns 9–11 of Table 1. Compounds of the compositions A_2BX_4 crystallize in different structures such as K_2SO_4, olivine, and spinel. Using the structure–chemical volume rule [4] and plotting the sum of the ionic radii $r_A + r_B + r_C$ as a function of the volume of the element, we obtain typical straight lines for the various structures. When this is done, it is found that $LaTi_2O_4$, $EuTi_2O_4$, $SrTi_2O_4$, and Eu_2TiO_4 are located on a line which represents the K_2NiF_4 structure. The lattice of K_2NiF_4 is shown in Fig. 4 [5]. In this lattice, each nickel ion is surrounded octahedrally by six fluorine ions. Four out of these six fluorine ions are shared with two nickel ions in order to form a two-dimensional lattice. The potassium and the other two fluorine ions lie in approximately the same plane, which is parallel to the basal plane. Since the lattice constants a and b differ slightly, it follows that the tetragonal unit cell of K_2NiF_4 has a slight orthorhombic distortion. $LaTiO_3$ crystallizes in a similar $CaTiO_3$-type lattice. $EuTiO_3$ has the same structure as the cubic $Me^{II}Ti^{IV}O_3$ perovskite but the unit cell is doubled.

Literature Cited

1. A. Meisel, M. Köstler, and A. Merkel, J. Prakt. Chem., 34:112 (1966).
2. M. A. Blokhin and A. T. Shuvaev, Izv. Akad. Nauk SSSR, Ser. Fiz., 26:429 (1962).
3. H. Holzapfel and J. Sieler, Z. Anorg. Allg. Chem., 343:174 (1966); J. Sieler and H. Hennig, J. Prakt. Chem., 34:168 (1966); J. Kaiser, Diploma Thesis, Leipzig (1966).
4. G. Gattow, Naturwiss., 50:152 (1963).
5. D. Balz and K. Plieth, Z. Elektrochem., 59:545 (1955).

DETERMINATION OF THE DEGREE OF IONICITY
OF THE BONDS IN SOME COMPOUNDS OF PHOSPHORUS
FROM THE SHIFT OF THE $K_{\alpha_{1,2}}$ X-RAY LINES*

É. P. Domashevskaya, Ya. A. Ugai, and O. Ya. Gukov

The degree of ionicity of the bonds in some compounds of phosphorus with zinc and cadmium was determined from the relative shifts of the $K_{\alpha_{1,2}}$ lines of phosphorus in the x-ray spectra of AlP, GaP, InP, ZnP_2, Zn_3P_2, CdP_2, Cd_3P_2, and P_2O_5. This method for the determination of the degree of ionicity is judged in relation to the proposed scheme of bonding in $A^{III}B^{V}$ semiconductors. The bond ionicities obtained were (in %): 30 for AlP, 20 for GaP and InP, 40 for ZnP_2, Zn_3P_2, and CdP_2, 20 for Cd_3P_2, and 50 for P_2O_5 (the error was 10%).

The chemical shifts of the energy levels and of the atomic spectra of various solids are being used successfully in determining the atomic valence, bond ionicity, coordination numbers of elements in compounds, etc.

The chemical shifts of the γ-ray resonance spectra can be found readily by means of the Mössbauer effect. Unfortunately, there are only very few isotopes which can be used in the Mössbauer effect and, therefore, this method can be employed in studies of the chemical bonding in a limited number of substances [1].

Recently, Swedish scientists have developed an absolute method for the determination of the energy levels of elements and their compounds from their β-ray spectra [2, 3], which can be applied successfully in determining the chemical shifts of the inner energy levels of elements.

The x-ray spectroscopic method for investigating the chemical bonding in compounds is free of these limitations. Therefore, there is a continually growing use of x-ray spectra in studies of many aspects of the chemical bonding in solids. One such problem is the determination of the degree of ionicity of bonds (the charge of an element in a compound) from the chemical shifts of the x-ray emission lines.

The problem of the determination of the degree of ionicity of bonds is particularly important because in many solids, such as semiconducting compounds and alloys, we encounter mixing of the ionic, the covalent, and the metallic bonding.

We determined the degree of ionicity of bonds in some compounds of phosphorus with zinc and cadmium by measuring the relative shifts of the $K_{\alpha_{1,2}}$ x-ray emission lines of

---

* "Crystals," pp. 155–160 (see page 3).

phosphorus in AlP, GaP, InP, ZnP_2, Zn_3P_2, CdP_2, Cd_3P_2, and P_2O_5. The $K_{\alpha_{1,2}}$ lines were recorded by the primary method using the first-order reflection from a $(10\bar{1}0)$ quartz crystal with a resolution of 4000. The minimum values of the current and the voltage in the x-ray tube were V = 10 kV and I = 2 mA. In the absence of undesirable transformations on the anode of the x-ray tube, we recorded the Debye diffraction patterns of each compound before and after its deposition on the anode. In this way, we recorded 3-5 spectrograms for each compound and each was subjected to photometric measurements of 2-3 points in a spectral line. The shifts of the $K_{\alpha_{1,2}}$ lines were determined to within ± 0.05 eV (Table 1). These shifts indicated that, in all the compounds of phosphorus with metals, the $K_{\alpha_{1,2}}$ lines shifted in the direction of long wavelengths, whereas the lines of phosphorus pentoxide shifted in the direction of short wavelengths. Earlier, Faessler [4] had measured the shifts of the $K_{\alpha_{1,2}}$ lines of phosphorus for AlP, GaP, and InP (Table 1). His values of the shifts agreed particularly well with ours in the case of GaP and InP. Our shift for AlP (−0.27 eV) was practically identical with −0.29 eV deduced from the secondary spectrum [5].

According to the ideas developed in [6, 7], the long-wavelength shifts of the $K_{\alpha_{1,2}}$ lines of phosphorus in the spectra of phosphides and diphosphides, relative to the same lines in the spectrum of elemental phosphorus, are due to a shift of the electron cloud away from the electropositive metal to the electronegative phosphorus. This results in an acquisition of a negative charge by phosphorus. Therefore, the effective charge of the phosphorus decreases and its $K_{\alpha_{1,2}}$ lines shift in the direction of longer wavelengths. It is evident from Table 1 that the $K_{\alpha_{1,2}}$ lines of phosphorus shift toward shorter wavelengths only in the case of phosphorus pentoxide. This is because in P_2O_5 the phosphorus is bound to the more electronegative oxygen atoms (the electronegativity of phosphorus is 2.1 and that of oxygen 3.5) and the oxygen attracts the valence electrons of the phosphorus. Therefore, the screening of the inner-shell electrons of phosphorus by its valence electron decreases, the effective charge of the phosphorus nucleus increases, and the $K_{\alpha_{1,2}}$ lines of phosphorus shift toward shorter wavelengths.

The bond ionicity in phosphides and diphosphides can be estimated as follows. We suggested earlier a tetrahedral covalent bond scheme for $A^{III}B^V$ compounds [8]. This scheme is illustrated here for the case of AlP (Fig. 1). We can see that each atom is surrounded by a complete electron octet. The fourth electron pair, which forms a covalent bond, is supplied by an atom of P in the form of unshared s^2 electrons. These electrons are represented by large black dots. The electrons belonging to the Al atoms are represented by crosses.

Thus, one out of four covalent bonds, formed by each of the Al and the P atoms, is of donor−acceptor origin, i.e., it is a coordinate bond. This bond differs from the usual covalent bond only in the

TABLE 1. Shifts of $K_{\alpha_{1,2}}$ Lines and Bond Ionicity of
Some Compounds of Phosphorus

Compound	Shift of $K_{\alpha_{1,2}}$ lines of P, eV		Bond ionicity I(%), our data	Szigeti's effective charge [11]	Forbidden band width ΔE, eV
	our data	other values [4, 5]			
AlP	−0.27	−0.17 (−0.29)	30	0.30	3
GaP	−0.18	−0.17	20	0.58	2.4
InP	−0.18	−0.15	20	0.60	1.34
ZnP_2	−0.36	—	40	—	2.05[19]
Zn_3P_2	−0.36	—	40	—	1.24 [18, 19]
CdP_2	−0.36	—	40	—	2.02 [20]
Cd_3P_2	−0.18	—	20	—	0.6 [18, 19]
P_2O_5	+0.45	—	50	—	—
Error	±0.05		±10%		

```
Al  :  P  :  Al  :  P  :  Al
 •+     +•     +•     +•     +•
 P  :  Al  :  P  :  Al  :  P
 •+     +•     +•     +•     +•
Al  :  P  :  Al  :  P  :  Al
 +•     +•     +•     +•     +•
 P  :  Al  :  P  :  Al  :  P
 +•     +•     +•     +•     +•
Al  :  P  :  Al  :  P  :  Al
```

Fig. 1. Tetrahedral covalent bonds in AlP.

origin of the bonding electron pair, and, in the final analysis, the coordinate and the covalent bonds are perfectly identical. Determinations of the electron density distribution have indicated that the lattice sites in tetrahedral structures are occupied by atomic cores joined by electron "bridges" consisting of valence electron pairs. These bridges can be more or less smeared out, depending on the average atomic number of the compound. This smearing represents metallization of the bonds. Moreover, such "bridges" are more or less symmetrical in respect of the electron density distribution, depending on the difference between the electronegativities of the atoms occupying the lattice sites. The degree of ionicity of the bond is governed by this difference: the ionicity increases with increasing difference between the electronegativities of the atoms forming a bond. In the limit, a purely ionic bond is characterized by the complete separation of the charge so that an electron pair forming a "bridge" is drawn fully toward the more electronegative partner.

In a purely ionic compound, the complete transfer of electron pairs from the A^{III} to the B^V atoms should give rise to triply charged positive ions A^{3+}. Therefore, estimating the bond ionicity in AlP, we shall assume that 100% ionicity represents the state in which aluminum is triply ionized (Al^{3+}). The conversion of neutral Al to Al^{3+} causes a certain shift of the $K_{\alpha_{1,2}}$ lines of aluminum, ΔE, which corresponds to the 100% bond ionicity. Therefore, the experimentally observed shift of the $K_{\alpha_{1,2}}$ lines of aluminum, ΔX_{Al} in the spectrum of AlP should be related linearly to the bond ionicity I of this compound. Thus, we can define the bond ionicity as

$$I = \Delta X_{Al}/\Delta E \cdot 100, \%. \tag{1}$$

In the determination of ΔE of Al, we can use the calculations concerned with Al atoms and Al^{3+} ions, carried out using the self-consistent field method making allowance for the exchange [9, 10]. These calculations give $\Delta E = 0.823$ eV. The experimental shift of the $K_{\alpha_{1,2}}$ lines of aluminum in AlP is $\Delta X_{Al} = 0.24$ eV. Therefore, Eq. (1) gives I = 30% for this compound. The other $(100 - I)\%$ represents the nonpolar (covalent and metallic) components of the bonding.

Knowing the bond ionicity of AlP ($I_{AlP} = 30\%$) and the shift of the $K_{\alpha_{1,2}}$ lines of phosphorus in the spectrum of this compound ($\Delta X_P = -0.27$ eV), we can use the linear law to determine the bond ionicity I for the other compounds, employing the experimentally obtained shifts of the $K_{\alpha_{1,2}}$ lines of phosphorus ΔX_P for these compounds.

The values of the effective Szigeti charges e_S are known for AlP, GaP, and InP [11].

These effective charges were introduced by Szigeti to obtain agreement between the theoretical and the experimental relationships between the characteristic frequencies of the vibrations of the atoms in the lattice and the permittivity [12, 13].

Tolpygo [14], who extended the Born theory of crystal lattices to the case of mixed ionic-covalent bonding, points out with justification that the effective Szigeti charges is not a physical quantity which characterizes a given compound: it is simply one of the characteristics of the limiting optical vibrations. This is why the effective charge of SiC is found to be larger than that of NaCl (0.94 and 0.75 electrons, respectively). This result is obtained because the

Szigeti charge does not represent the deviation of the nature of the bonds from purely ionic, which is the starting point in the Born theory, but the influence of the approximations made in this theory.

Our estimates of the bond ionicity I showed that, among the $A^{III}B^V$ compounds listed in Table 1, AlP has the highest ionicity (I = 30%) and GaP has the same ionicity (I = 20%) as InP.

According to current ideas, an increase in the bond ionicity is accompanied by broadening of the forbidden band because of an increase in the difference between the maximal and the minimal values of the potential in a crystal. This difference decreases with increasing covalence of the bonds because the dips between atoms are filled by electron "bridges" and, therefore, the forbidden band width decreases (this width is proportional to the depth of the potential wells in a crystal).

Our results are not in conflict with these ideas. AlP exhibits the highest ionicity among the $A^{III}B^V$ phosphides listed in Table 1; it also has the widest forbidden band. According to our data, the ionicity of GaP and InP is the same but their forbidden band widths are not. The narrower forbidden band of InP, compared with the other $A^{III}B^V$ phosphides, is due to the stronger metallization of the chemical bonds in this compound.

In this connection, it is worth mentioning the role of the metallic bonding, which is one of the nonpolar components of the remaining $(100 - I)\%$ of the bonding in a semiconductor. X-ray diffraction studies of the distribution of the electron density in metals [15] indicate that free electrons are distributed throughout the whole volume of a crystal. Such a distribution of free electrons fills the dips in the potential more effectively than do the electron "bridges" of the covalent bonds. Therefore, the forbidden band should become narrower, decreasing to zero in the case of a pure metal.

The measured values of the shifts indicate that the bond ionicity in $A^{II}B^V$ compounds is 40% in the case of ZnP_2, Zn_3P_2, CdP_2, and 20% in the case of Cd_3P_2.

Relative measurements of the enthalpy of the formation of Zn_3P_2 [16] suggest that this compound is predominantly ionic. According to our data, 60% of the bonding in this compound is of the covalent-metallic type and the ionic component represents only 40%. Our view is supported also by the measurements of the forbidden band width of this compound, which is of the order 1.2 eV [17, 18] and does not exceed the values typical of $A^{III}B^V$ semiconductors with predominantly covalent-metallic bonds.

According to our data, the bond ionicity of Cd_3P_2 (20%) is less than that of Zn_3P_2, which is in good agreement with the values of the forbidden band widths ΔE of these compounds (Table 1).

The values of the bond ionicity can be used to predict the forbidden band width of zinc and cadmium diphosphides: this width should be 1.5-2 eV. This prediction is supported by the recent [19, 20] measurements of the optical properties of these compounds which yield the optical width of the forbidden band $\Delta E = 2$ eV for ZnP_2 [19] and CdP_2 [20].

Literature Cited

1.   G. K. Wertheim, Mössbauer Effect: Principles and Applications, Academic Press, New York (1964).
2.   C. Nordling, E. Sokolowski, and K. Siegbahn, Ark. Fys., 13:483 (1958).
3.   K. Siegbahn et al., Atomic, Molecular and Solid State Structure Studied by Means of Electron Spectroscopy, Uppsala (1967).

4. A. Faessler, Proc. Fifth Intern. Conf. on Physics of Semiconductors, Prague, 1960, publ. by Academic Press, New York (1961), p. 914.

5. N. Johnson, Dissertation, Lund (1935).

6. S. M. Karal'nik, Izv. Akad. Nauk SSSR, Ser. Fiz., 20:818 (1956).

7. A. T. Shuvaev, Dissertation for Candidate's Degree [in Russian], Rostov-on-Don (1964).

8. Ya. A. Ugai and É. P. Domashevskaya, Dokl. Akad. Nauk SSSR, 156:430 (1964).

9. C. Froese, Proc. Cambridge Phil. Soc., 53:206 (1957).

10. E. Clementi, A. D. McLean, D. L. Raimondi, and M. Yoshimine, Phys. Rev., 133:A1274 (1964).

11. D. A. Kleinman and W. G. Spitzer, Phys. Rev., 118:110 (1960).

12. B. Szigeti, Trans. Faraday Soc., 45:155 (1949).

13. B. Szigeti, Proc. Roy. Soc., London, A204:51 (1950).

14. K. B. Tolpygo, in: Chemical Bonds in Semiconductors and Solids, Consultants Bureau, New York (1967), pp. 121, 129.

15. N. V. Ageev and D. L. Ageeva, Izv. Akad. Nauk SSSR, Otd. Khim. Nauk, No. 1, p. 17 (1948).

16. G. V. Samsonov and L. L. Vereikina, Phosphides [in Russian], Kiev (1961).

17. M. V. Vol'kenshtein, Structure and Physical Properties of Molecules [in Russian], Izd. AN SSSR, Moscow (1955).

18. H. J. Yearian, Phys. Rev., 48:631 (1935).

19. I. J. Hegyi, E. E. Loebner, E. W. Poor, Jr., and J. G. White, J. Phys. Chem. Solids, 24: 333 (1963).

20. W. Zdanowicz and A. Wojakowski, Phys. Status Solidi, 10:K93 (1965).

DERIVATION OF THE ELECTRONIC STRUCTURE OF ALUMINUM ATOMS IN Al$_2$O$_3$ FROM X-RAY SPECTROSCOPIC DATA*

S. M. Karal'nik and É. P. Domashevskaya

An allowance for the change in the external screening of the inner-shell electrons of aluminum atoms and a numerical estimate of this change made it possible to explain the observed chemical shift of the x-ray K lines and of the absorption edge of aluminum resulting from the oxidation of this metal. It was found that this change involved not only total transfer of the outer electrons from the aluminum to the oxygen atoms but also the partial excitation of the aluminum atoms.

The x-ray spectrum of an element and its change resulting from the formation of a compound by the element can be used to obtain information on the change in the electronic structure due to the formation of a compound. The principal experimental data on the changes in the x-ray spectrum of aluminum which are caused by the formation of Al$_2$O$_3$ can be summarized as follows [1]:

1) the K$_{\alpha_{1,2}}$ lines of aluminum shift toward shorter wavelengths by an amount $\sim$ 2 X.U., i.e., by $\sim$ 0.35 eV;
2) the K$_\beta$ line shifts toward longer wavelengths by an amount 21−27 X.U. (the values reported differ from one author to another) or by 4.1−5.3 eV;
3) the ratio of the intensities of the K$_\beta$ to the K$_a$ lines increases by 18%;
4) the K absorption edge of aluminum shifts toward shorter wavelengths by about 5.5 eV;
5) the L$_{II,III}$ absorption edge also shifts toward shorter wavelengths (higher energies) by 3.7 eV [2].

According to one of the present authors and Ugai [3], the K$_{\alpha_{1,2}}$ lines of aluminum in Al$_2$O$_3$ shift toward shorter wavelengths by +0.48 eV and the maximum of the K$_{\beta_1}$ band of aluminum in Al$_2$O$_3$ shifts toward longer wavelengths by −4.85 eV relative to the maximum of the K$_{\beta_x}$ band of metallic aluminum (this was deduced from the K$_\beta$ spectra corrected by the column method for the distortions introduced by the apparatus).

The changes in the K series are shown schematically in Fig. 1.

We shall attempt to interpret the observed changes using an approach in which an allowance is made for the changes in the external screening of the inner by the outer electrons due to alterations in the chemical bonding [4, 5]. It is assumed that a shift in the energy positions of

---

* "Crystals," pp. 161-166 (see page 3).

the inner levels is due to a change in the screening resulting from the participation of additional outer electrons in the formation of bonds. The change in the energy of the K or L level due to external screening by a charge σ will be assumed to be given by $\Delta E = \Delta E_{z, z+1}\, \sigma$, where $\Delta E_{z, z+1}$ is the difference between the energies of the K (or L) level for elements z and z + 1, which corresponds to a shift of the level resulting from the reduction of the nuclear charge by unity.

The external screening can be determined only if we know the probability of finding the relevant outer (valence) electrons within the inner shells of an atom.

The external screening of the inner electrons by the valence electrons of charge σ can be represented by the electronic density of a valence electron of suitable symmetry located within an inner orbit nl, i.e., by $\sigma = \int_0^r P^2_{n'l'}(r)\,dr$; here, $P_{n'l'}$ is the radial wave function of the valence electron.

The external screening of the 1s and 2p inner electrons by the 3p or 3s valence electrons can be found by calculating the integrals $\int P^2_{3p}(r)\,dr$ and $\int P^2_{3s}(r)\,dr$ between zero and the value of the radius corresponding to the maximum of the wave function of the 1s or 2p inner electrons.

We carried out this integration by a numerical method [6]. The calculation of the screening integrals required a knowledge of the wave functions P_{3p} and P_{3s} which could be found most accurately by solving the wave equation by the self-consistent field method. Since the values of the wave functions P_{3p} and P_{3s} of the aluminum atom were not available in the numerical form, we had to make our calculations ignoring the exchange interaction. We used the state of Al$^+$, calculated by the self-consistent field method [7], as the zeroth approximation. Since the contribution of the 3p valence electron of Al to the potential in the region of localization of the 1s and 2p electrons was very small, the changes in the wave functions ΔP_i associated with the transition from Al$^+$ to Al were calculated in the first approximation of the perturbation theory on the basis of self-consistent changes in the wave functions and the potential.

The probability of finding a 3p electron within the 1s shell was thus found to be $\int_{r=0}^{r=0,08} P^2_{3p}(r)\,dr = 0,00005$ (r = 0.08 corresponds to the maximum of the P_{1s} function) and the corresponding probability for a 3s valence electron was $\int_0^{r=0.08} P^2_{3s}(r)\,dr = 0.0025$. The same 3p and 3s electrons screened much more strongly than the 2p shell:

$$\int_{r=0}^{r=0.40} P^2_{3p}(r)\,dr = 0.0085 \quad \text{and} \quad \int_{r=0}^{r=0.40} P^2_{3s}(r)\,dr = 0.0220$$

(r = 0.40 corresponds to the maximum of the P_{2p} function).

When the bonding results in the loss of a 3s electron, the energy of the 1s level should increase in proportion to the external screening of the 1s shell by the 3s electron. Thus, if the energy of the 1s level of aluminum is 1560 eV and the loss of one K electron is equivalent to an effective increase in the nuclear charge by unity, i.e., to a transition to the next element (silicon) in the periodic table whose 1s level energy is 1842 eV, we find that the removal of only 0.0025 of an electron from the vicinity of a nucleus should also increase the nuclear charge and this increase should shift the K level by $0.0025 \cdot (1842 - 1560) \approx 0.7$ eV. Since the oxidation of aluminum Al $\rightarrow$ Al$_2$O$_3$ results in the loss of both 3s electrons, the shift of the 1s level should be at least double (it is likely to be more than double because the external screening caused by the second 3s electron in the absence of the first is stronger owing to the stronger binding to the atomic core), i.e., the total shift of the 1s level should be close to 1.5 eV.

Since the external screening of the 1s shell by the 3p electron is small, the loss of the 3p electron resulting from the oxidation of aluminum should have very little influence on the shift of the K level.

When the same 3s electrons are lost by aluminum in the bonding process, we find that the shift of the 2p level ($L_{II, III}$) should be $2 \cdot 0.02 \cdot (103 - 73) \approx 1.3$ eV. This estimate shows that the inner K and L levels shift by approximately the same amount in the $Al \rightarrow Al_2O_3$ reaction. However, the K level "leads" the L level in respect of the shift and, therefore, the $K_{\alpha_{1,2}}$ line should be displaced toward shorter wavelengths by ~ 0.2 eV, which is very close to the experimental value of the shift.

This estimate ignores the contribution of the shift of the 2p level due to the loss of the 3p electron, which represents ~ 0.35 of the contribution of the 3s electron (0.008 and 0.022). This has been done deliberately because an increase in the intensity of the K_β line in Al_2O_3 (point 3 in our summary) indicates clearly that the 3p shell of aluminum does not become vacant after the $Al \rightarrow Al_2O_3$ reaction (otherwise the K_β line would not have been observed at all for Al_2O_3). In fact, the 3p shell becomes partly occupied by the 3s electrons.

The shift of the K_β line toward longer wavelengths (point 2 in our summary) can be explained if we can demonstrate that the initial level ($M_{II, III}$) of the transition corresponding to this line shifts in the same direction as the K level but the shift of the M level is greater because of the loss of the outer 3s and 3p electrons from the aluminum to bonds with oxygen.

This explanation is correct because two 3s electrons are removed from the M_I shell, which is equivalent − in the case of the $M_{II, III}$ level − to a transition to an element (phosphorus) whose atomic number is two units higher. The energy of the $M_{II, III}$ level of phosphorus is 6.4 eV, whereas the corresponding energy of aluminum is 2.7 eV [8]. Therefore, this level shifts by $6.4 - 2.7 = 3.7$ eV, i.e., the 3p level shifts ahead of the 1s level by $3.7 - 1.5 \approx 2$ eV. This means that the K_β line should shift by ~ 2 eV toward longer wavelengths because the energy gap between the 1s and 3p levels decreases. This result is in qualitative agreement with the experimental data and the discrepancy of 2 eV is not surprising in view of the considerable width of the K_β line and the known indeterminacy of its position and profile.

The following comments can be made about the shifts of the K and L absorption edges of aluminum. Our calculations predict that the contribution of the shift of the inner levels should be about 1.5 eV in both cases. This value disagrees with the experimental data because of the indeterminacy of the absorption edge positions and because of some underestimation of the shifts of the K and L levels. Moreover, our description of the changes in the electronic structure of aluminum, based on the assumption that the outer electrons are displaced away from aluminum

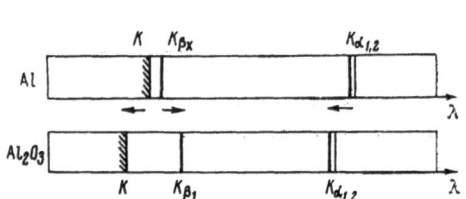

Fig. 1. Schematic representation of the positions of the K x-ray emission lines and absorption edges of metallic aluminum and its oxide. The diagram is based on the experimental data reported in [1, 3].

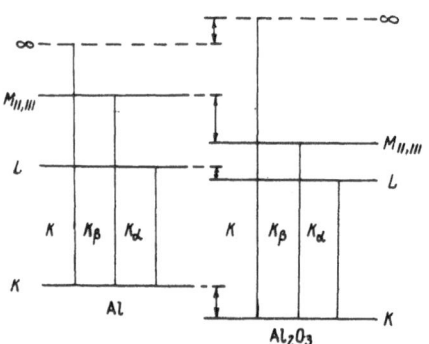

Fig. 2. Schematic representation of the changes in the energy levels resulting from the oxidation of metallic aluminum to Al_2O_3.

to form chemical bonds but are not lost completely to oxygen, should be modified to allow for the fact that the inner K electron which absorbs an x-ray quantum, should be raised not only from a greater "depth" (because of the shift of the K level) but also to a greater "height." This increases the energy necessary for the detachment of the K electron and, therefore, the shift of the K edge toward higher energies is greater than the shift of the K level. In other words, the cause of the shift of the K or L edge lies in the lower position of the initial level and the higher position of the final level in the absorption event.

Figure 2 shows schematically the energy levels and transitions deduced from our calculations.

Our atomic approach to the changes in the x-ray spectra (used also by other investigators [9]) postulates that the increase in the energy gap between the K edge and the K$_\beta$ line (more accurately, the short-wavelength component of this line) reflects an increase in the degree of participation of the outer electrons in the bonds, which reduces the probability of the participation of these electrons in the conduction process. This effect is to be expected when a metal is converted to an oxide, which is a semiconductor or an insulator. In the energy band model, this is represented by an increase in the energy gap (forbidden band).

Thus, the observed changes in the x-ray spectra can be explained by changes in the external screening of the electrons because of the different nature and degree of atomic bonding. We may conclude that the Al → Al$_2$O$_3$ oxidation process does not result in a complete transfer of the outer electrons from aluminum to oxygen because these electrons are still coupled appreciably in their parent atoms, i.e., we find that the bonds in aluminum oxide are of the mixed covalent-ionic type [3].

Literature Cited

1.   A. Meisel, Phys. Status Solidi, 10:365 (1965).
2.   V. A. Fomichev, Fiz. Tverd. Tela, 8:2892 (1966).
3.   É. P. Domashevskaya and Ya. A. Ugai, Röntgenspektren und chemische Bindung, Leipzig (1966), p. 70.
4.   S. M. Karal'nik, Izv. Akad. Nauk SSSR, Ser. Fiz., 20:815 (1956).
5.   S. M. Karal'nik, Izv. Akad. Nauk SSSR, Ser. Fiz., 21:1445 (1957).
6.   D. R. Hartree, The Calculation of Atomic Structures, Wiley, New York (1957).
7.   L. Bierman and H. Harting, Z. Astrophys., 22:81 (1943).
8.   M. A. Blokhin (ed.), X-Rays [Russian translation], IL, Moscow (1960), p. 284.
9.   B. N. Das and L. V. Azaroff, Acta Met., 13:827 (1965).

INVESTIGATION OF THE ATOMIC BINDING FORCES IN MERCURY TELLURIDE BY THE X-RAY DIFFRACTION METHOD*

L. I. Kleshchinskii, P. V. Sharavskii, and A. I. Inyutkin

An x-ray diffraction examination was made of mercury telluride crystals prepared by various methods. Measurements of the intensities of the superstructure maxima and calculations of the characteristic temperatures and of the lattice period were used to obtain information on the nature of the dependence of the atomic forces on the concentration of mercury vacancies.

Mercury telluride has the sphalerite structure and is homogeneous over a wide range of nonstoichiometric concentrations [1, 2].

Stoichiometric HgTe is very difficult to prepare because of the volatility of mercury. The various techniques used in the attempts to prepare stoichiometric HgTe frequently yield defective samples. This is why the parameters of this substance reported by different authors are frequently contradictory.

Variations in the degree of nonstoichiometry naturally lead to changes in the atomic binding forces and in the integrated intensities of the selective x-ray reflection. A comparison between the calculated and the measured values of the integrated reflection intensities is the only (and very reliable) method for proving the perfection of a crystal.

Our preliminary investigations of mercury telluride crystals grown from a mercury-rich melt [3] demonstrated the need for a considerable improvement in the sensitivity of the available x-ray diffraction apparatus. With this point in mind, we assembled a scintillation attachment for the detection of x rays. This attachment included the transistor circuit described in [4].

A germanium single crystal, oriented along the (111) plane, was used as a monochromator. The careful selection of a germanium crystal with a high reflectivity and improvements in the factory-built monochromator unit increased the intensity by a factor of 2.0−2.5.

We found that an appreciable increase in the sensitivity of the apparatus could be achieved only if we used high-quality scintillation crystals and photomultipliers, and if we ensured satisfactory conditions for the collection of light by the multiplier photocathode. Moreover, a satisfactory power supply had to be provided for the FÉU-35 photomultiplier and good-

---

* "Crystals," pp. 167−172 (see page 3).

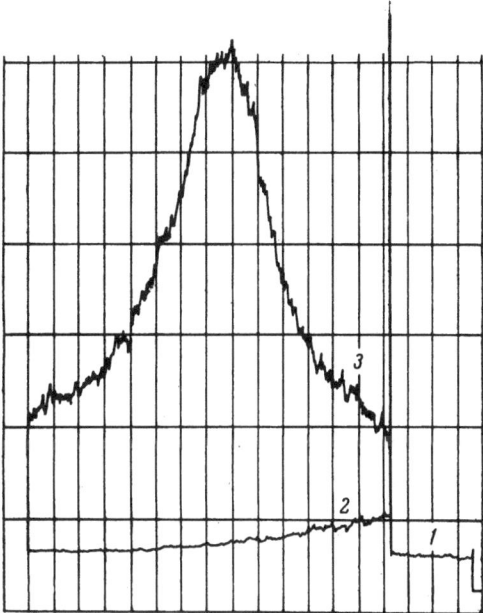

Fig. 1. Level of the intrinsic noise of
the recording system combined with the
cosmic component of the background (1)
and the effect of the scattering by air (2).
Curve 3 represents the diffuse background
maximum of liquid mercury.

quality transistors had to be employed. The whole recording system had to be stabilized (this was achieved by providing a battery power supply). Our apparatus had to satisfy stringent requirements, primarily because the sensitivity in the case of weak lines depended on the ratio of the signal-to-noise amplitudes. The maximum gain in the signal amplitude could be achieved only by optimal utilization of the intensity of the light flashes in the scintillator. This was ensured by introducing a thin layer of silicone adhesive between the scintillator and the photocathode. Photomultipliers of the FÉU-35 type exhibited a considerable scatter of their parameters even within the same batch. The photomultiplier was selected by trial and error, bearing in mind the requirements of uniform surface distribution of the photocathode sensitivity and minimum level of intrinsic noise. Next, we selected the resistance of the voltage divider to guarantee the maximum pulse amplitude and the optimum ratio of this amplitude to the noise level. This was necessary to ensure full utilization of the good time resolution of the photomultiplier.

These improvements and the matching of the time resolution throughout the recording channel enabled us to determine the integrated intensity down to the level of 2–4 electrons and to record the diffuse background directly on a potentiometer chart. This made it possible to reveal fine details of the structure.

The samples used in our investigation were prepared by different methods. Thus, samples 1–4 were the purest and were prepared by the floating molten zone method. Samples 5–6 were grown by the Bridgman method.[1] Sample 7 was prepared by the Bridgman method from a melt containing 25 at.% of excess mercury.

---

[1] HgTe samples 1–6 were kindly supplied by V. I. Ivanov-Omskii and K. P. Smekalova of the Leningrad Physicotechnical Institute.

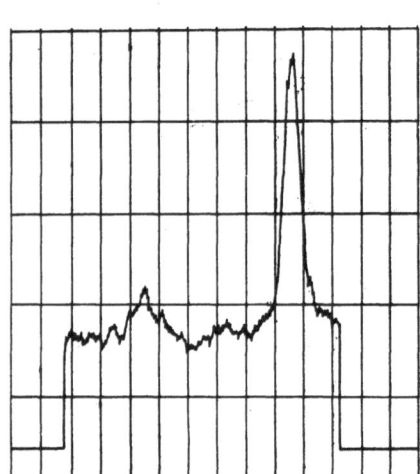

Fig. 2. Signal of the (200) selective reflection of HgTe and the diffuse background maximum for a sample (No. 7) prepared from a mercury-rich melt.

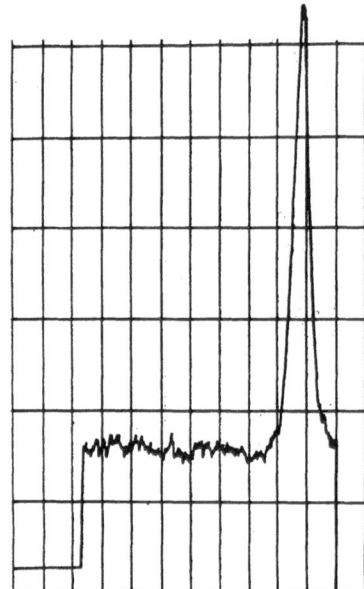

Fig. 3. Part of the x-ray diffraction pattern of HgTe sample 1, recorded between $2\theta = 25°$ and $2\theta = 36°$.

All these samples, except No. 7, were powders of 5-7 μ grain size, which had been elutriated in toluene. Sample 7 was prepared by simple grinding in an agate mortar (this sample was used simply to check the presence of a second phase in a method for growing stoichiometric crystals from a melt which was enriched with mercury in order to establish an equilibrium vapor pressure during growth).

A maximum of the diffuse background was found in the spectrum of sample 7 (Fig. 2) on the left of the (200) selective reflection. This maximum was attributed to the presence of a second phase since it was not observed in the samples prepared by the floating molten zone method or by the Bridgman method (Fig. 3).

Figure 1 shows the diffuse background maximum of liquid mercury. The diffuse background maxima of Figs. 1 and 2 were observed at the same angle. This indicated that the method of preparing stoichiometric samples, in which the vapor pressure of mercury was regulated by varying the amount of excess mercury in the melt, could give rise to defective samples, which would have poor electrical properties (especially at low temperatures).

The x-ray diffraction patterns of samples 1-6 were used to determine the lattice period, the characteristic temperatures, and the integrated reflection intensities. These intensities were then employed to find the deviations from stoichiometry. The lattice period was measured using monochromatic Fe K_α radiation. The angles were read directly from the goniometer scale.

The characteristic temperatures and the integrated intensities of all the reflections were determined using monochromatic Cu K_α radiation.

The characteristic temperatures Θ were deduced by a graphical method [5] from the integrated intensities at T = 293 and 77°K and from the thermal component of the diffuse background. The measured background intensity was converted to the absolute values by means of

TABLE 1. Electrical Properties, Squares
of Structure Amplitudes F^2, and Characteristic Temperatures Θ

Sample No.	R, cm^3/C	$\sigma, \Omega^{-1} \cdot$ cm^{-1}	F_{200}^2 (calc.)	F_{200}^2 (exper.)	Θ, °K
1	−7500	128	542	500 ± 4	130 ± 5
2	−6860	88	542	500 ± 4	130 ± 5
3	−6820	53	542	500 ± 4	130 ± 5
4	− 97	19	542	490 ± 4	120 ± 5
5	− 20	30	542	480 ± 4	110 ± 5
6	− 20	30	542	480 ± 4	110 ± 5

a fused quartz standard. The cosmic component of the background, the scattering by air, and the intrinsic noise level of the recording system were determined experimentally (Fig. 1) and deduced from the total background. The contribution of the Compton scattering was calculated theoretically. The double Bragg scattering was ignored because its contribution was small. The Debye function and the Lorentz polarization factor were calculated on a Razdan computer. The atomic scattering factors f_{Hg} and f_{Te} were determined for neutral atoms, making correction for the anomalous dispersion [6]. The values of F^2 were calculated by converting the selective reflection intensities into the absolute values by means of an NaCl standard.

It is evident from Table 1 that the Hall coefficient R and the electrical conductivity σ, measured at 1.7°K, were the highest for samples 1–3. These samples were characterized also by the largest values of the characteristic temperatures. An analysis of the atomic scattering factors f_{Hg} of samples 5 and 6 indicated that they were 0.44% smaller than the factors f_{Hg} for samples 1–3. This indicated that samples 5 and 6 were not stoichiometric but deficient in mercury. The lattice period was the same for all the samples and equal to 6.4590 ± 0.0005 Å. The constancy of the lattice period could be explained by the superposition of two effects. The formation of vacancies at the expense of the component with the larger atomic radius reduced the lattice period but the weakening of the atomic binding forces compensated this reduction. This was confirmed by a decrease in the characteristic temperatures of samples 5 and 6.

Thus, our analysis of the experimental results indicated that samples 1–3 were the closest to stoichiometric composition. However, this conclusion should be checked because of the discrepancy between the experimental and the theoretical structure amplitudes, which could be due to a considerable contribution of the ionicity to the bonding in HgTe, or to a deviation from the stoichiometric composition (a deficiency of mercury in the crystal lattice could reduce the integrated intensities). We used the method described in [7] to determine the effective charges on the ions.

The effective charge deduced from our x-ray diffraction data was 0.60 ± 0.05 electron. The effective charge was found using the Szigeti relationship given for HgTe in [8], where an effective ionic charge of 0.6 electron was obtained. This result supported our hypothesis of the stoichiometry of samples 1–3 and also indicated the degree of ionicity in the bonding of atoms in mercury telluride.

Literature Cited

1. R. T. Delves and B. Lewis, J. Phys. Chem. Solids, 24:549 (1963).
2. R. T. Delves, Brit. J. Appl. Phys., 16:343 (1965).
3. A. I. Inyutkin and L. I. Kleshchinskii, in: Papers Presented 23rd Conf. at Leningrad. Structural Engineering Institute, Physics Section [in Russian], Leningrad (1965), p. 27.

4.   E. A. Volosovichute, A. I. Inyutkin, L. I. Kleshchinskii, and P. V. Sharavskii, in: Papers Presented at 24th Conf. at Leningrad Structural Engineering Institute, Physics Section [in Russian], Leningrad (1966), p. 12.
5.   V. I. Stafeev and P. N. Aronov, Kristallografiya, 4:85 (1959).
6.   D. T. Cromer and J. T. Waber, Acta Crystallogr., 18:104 (1965).
7.   N. N. Sirota and E. M. Gololobov, Dokl. Akad. Nauk SSSR, 156:1075 (1961).
8.   D. H. Dickey and J. G. Mavroides, Solid State Commun., 2:213 (1964).

INFLUENCE OF THE CHEMICAL BONDING ON THE $K_{\alpha_{1,2}}$ DOUBLET OF IRON INVESTIGATED USING LOCAL X-RAY SPECTROSCOPIC ANALYSIS*

G. N. Ronami, A. K. Milai, and L. I. Tutova

The method of local x-ray spectroscopic analysis (the microanalyzer method) is used in investigations of the chemical bonding to identify small amounts of various phases and inclusions in many branches of physics, chemistry, metallurgy, geology, etc. Some methodological aspects of this method are considered in the present paper.

Investigations of the possibility of using local x-ray spectroscopic analysis in the identification of the state of elements in microscopic regions have been started in the X-Ray Spectroscopy Laboratory in the Physics Department of the Moscow State University.

X-ray emission spectra are sufficiently sensitive to changes in the density of electron states to provide useful information in studies of the nature of chemical bonding. The relationships between the x-ray emission spectra and the state of the elements in compounds and alloys (and, consequently, the physical properties of matter) have already been established. Local x-ray spectroscopy can be used to tackle many special cases. The microanalyzer (MS-46) used in these investigations is capable of dealing with the emission spectra of elements ranging from boron to uranium. The analysis can be limited to a region of $1\,\mu$ diameter.

One of the many applications of local x-ray spectroscopy is the determination of the state of the elements in geological samples. Such samples cannot be tackled by other methods because they consist of many phases and the dimensions of a given phase may be only a few hundredths of a micron. Only the microanalyzer method can be used to determine the quantitative composition and the state of the elements in such small regions.

The object of our study was the identification of the state of iron in microscopic regions in natural minerals such as oxides and sulfides.

According to the available information, the compounds of iron and sulfur which we investigated should contain iron in the divalent and trivalent states. Our task was to confirm this information and to determine the proportions of divalent and trivalent ions in these compounds. We began our study by investigating the capabilities of the MS-46 microanalyzer in x-ray emission spectroscopy. This was done by investigating the K_{α_1} and K_{α_2} lines of iron in Fe_2O_3

---

* "Crystals," pp. 173–176 (see page 3).

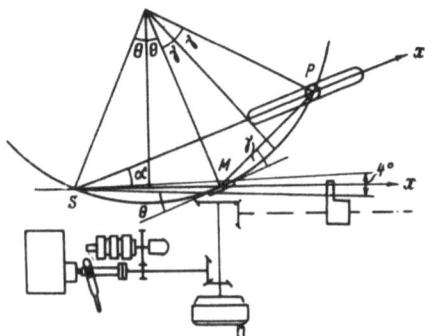

Fig. 1. Basic layout of the spectrometer:
M, analyzer crystal; S, x-ray source; P,
counter; $\alpha = 20°30'$; Θ, angle of incidence
of the x rays.

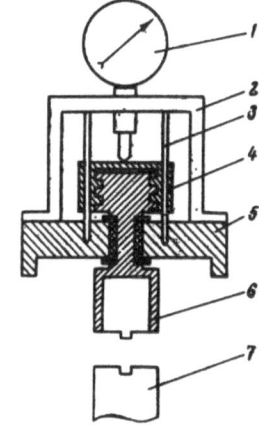

Fig. 2. Basic layout of the attachment used
in recording x-ray spectra: 1) micrometer
gauge; 2) gauge support; 3) guiding pillars;
4) double-threaded nut; 5) casing; 6) double-
threaded drive shaft; 7) coupling.

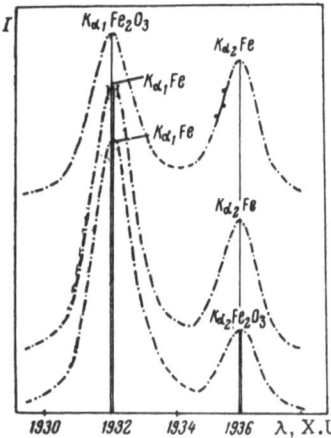

Fig. 3. K_{α_1} and K_{α_2} lines of iron in Fe_2O_3
and the $K_{\alpha_{1,2}}$ doublet of pure iron.

and in pure iron because these lines were least affected by the influence of chemical bonding.
A quantitative x-ray spectroscopic analysis showed that the composition of Fe_2O_3 was nearly
stoichiometric. The analyzer was a crystal of quartz of the (1010) orientation with an interplanar distance d = 4.2459 Å, bent to a radius of 250 mm. The spectrum was recorded in the
third reflection order. The dispersion of the spectrometer was 5.66 X.U./mm. The basic layout of the spectrometer is shown in Fig. 1.

The intensities in the x-ray spectrum were recorded with a gas-flow proportional counter
at intervals of 0.07−0.05 X.U. A special attachment was used to cover the spectrum at these
intervals (Fig. 2). Damage to the surface of a sample was avoided by moving it slowly (the
electron probe was kept in a fixed position).

TABLE 1. Shifts of the $K_{\alpha_{1,2}}$ Lines of Iron in Fe_2O_3
Relative to Pure Element

Lines	$\Delta\lambda$, X.U.	$\delta\lambda$, X.U.	Shift ΔE, eV
$K_{\alpha_2}Fe - K_{\alpha_1}Fe$	3.93 ± 0.02	—	—
$K_{\alpha_2}Fe_2O_3 - K_{\alpha_1}Fe$	3.98 ± 0.02	$+0.05$	-0.18 ± 0.05
$K_{\alpha_2}Fe - K_{\alpha_1}Fe_2O_3$	3.99 ± 0.02	-0.06	$+0.21 \pm 0.05$

Figure 3 shows the K_{α_1} and K_{α_2} lines of iron in Fe_2O_3 and the $K_{\alpha_{1,2}}$ doublet of pure iron. Each curve in Fig. 3 is the average of 5–7 experimentally obtained curves. Each point in this figure represents, on the average, $(4-7) \cdot 10^4$ pulses.

Pure iron was used as the standard in recording the K_{α_1} and K_{α_2} lines. The shift of the K_{α_1} line of Fe in Fe_2O_3, relative to its position in pure iron, was determined by measuring the position of the K_{α_1} line of iron in the oxide and of the K_{α_2} line of iron in the pure element. The distance between the maxima of these lines was compared with the doublet separation for pure iron. The change in this distance indicated that the K_{α_1} line of iron in the oxide shifted relative to the corresponding line of the pure element. Similarly, the shift of the K_{α_2} line of iron in the oxide was measured using the K_{α_1} line of pure iron as the standard.

These measurements enabled us to find the shifts of the K_{α_1} and K_{α_2} lines of iron in the oxide relative to the corresponding lines of the pure element. The shifts of the two lines were of the order of 0.2 eV. The K_{α_1} line shifted toward shorter wavelengths and the K_{α_2} line shifted toward longer wavelengths (Table 1). This was not in conflict with the current ideas on the changes in the positions of the atomic energy levels resulting from the conversion of a pure transition metal to a compound.

Our ability to measure small shifts indicates that the microanalyzer method can be used to investigate the influence of chemical bonding on the x-ray emission spectra of microscopic regions by means of the local x-ray spectroscopic analysis.

INVESTIGATION OF THE CARBON K AND METAL EMISSION BANDS AND BONDING FOR STOICHIOMETRIC AND NONSTOICHIOMETRIC CARBIDES*

J. E. Holliday

Edgar C. Bain Laboratory
For Fundamental Research
United States Steel Corporation
Research Center
Monroeville, Pennsylvania

The carbon K and metal L and M emission bands from stoichiometric and nonstoichiometric transition metal carbides were measured with a high-resolving-power grating spectrometer. It was found possible to relate the shape of the C K band to bonding. Groups IV and V transition metal carbides, which are strongly bonded carbides, have narrow C K bands. For the weakly bonded carbides, Group VI and higher, the C K bands are broad and have a shape close to that of the C K band from graphite. The C K bands from MC_x carbides show an increase in asymmetry with decreasing values of x.

Introduction

Although the electronic properties and structure of transition metal carbides have been investigated extensively by other techniques such as magnetic susceptibility, resistivity, thermoelectric power, Hall effect, and x-ray diffraction, soft x-ray spectroscopy affords another important experimental method for studying bonding and electron distribution in these carbides. The technique used and the results of this study will also be of importance to the user of the electron microprobe interested in the detection and quantitative analysis of carbon. The method of investigation was to measure the change in shape and wavelength of both the carbon K and metal emission bands from transition metal carbides when a given series is crossed as well as band changes due to departure from stoichiometric composition. The changes in band shape were correlated with changes in the heats of formation and melting temperature. Some preliminary results of these studies have already been published [1].

Experimental Method

The electron transitions investigated in this paper were the carbon K emission band

---

* "Crystals," pp. 177-197 (see page 3).

(2p → 1s transition) and the metal $L_{II, III}$ (3d + 4s → 2p transition) and M_V emission bands (5p → 3d transition). All of the above transitions are from either conduction or valence bands and cover the wavelength region from 20 Å to 80 Å. The carbon K and the metal $L_{II, III}$ emission bands were measured with a curved grating spectrometer utilizing a platinum grating (ruled by Bausch and Lomb Co.) having a 1° blaze, 3600 grooves/mm, and a radius of curvature of 1 m. This grating was able to resolve the C K bands to a much greater extent than the C K bands previously reported [1] using the 2160 grooves/mm grating. The M_V bands were measured with the 2160 grooves/mm grating since these bands had insufficient intensity and P/B to be observed with the 3600 grooves/mm grating. The detector in the present experiments was a flow proportional counter using P-10 gas at a pressure of 8 cm of Hg. The beam current was 1.4 mA at a potential of 4000 V and the x-ray chamber was evacuated to $5 \cdot 10^{-8}$ torr. Complete details of the spectrometer have been described previously [2-4]. These conditions were in marked contrast to those of earlier investigators who measured C K bands with a grating [5-9] using photographic film for a detector, poor vacuum conditions, and extremely high power input (50-100 mA beam current).

Since most of the C K bands from carbides have a width at half maximum intensity ($W_{1/2}$) narrower than those from graphite and lampblack, trace amounts of free carbon on the carbide surface will distort the shape of the C K band. Thus, the surface must be thoroughly cleaned and there must be no carbonaceous contamination deposited by the electron beam during the course of the measurements. In addition, coating the surface of nonconducting carbides and diamond with carbon to prevent charging due to the electron beam will also result in a distortion of the true shape of the C K band from the material under investigation. Since it was previously reported [4] that the rate of carbon contamination is directly related to the rate of degassing of the target, the targets were first thoroughly degassed and then cleaned by ion bombardment before the spectra were measured. It should be pointed out that the increase in the peak intensity of the C K band from the carbides during electron bombardment cannot be used as a test for carbon contamination. This is because the peak wavelengths of the carbide C K bands are shifted between 0.2 Å and 0.34 Å relative to the peak wavelength of the graphite C K band (44.85 Å). The peak shift in the present experiments is defined as the wavelength difference between the maximum intensities of the two bands under consideration. The test used in the present measurements was the constancy of the shape of the C K band. If after several runs and a number of ion cleanings, the shape of the C K band was reproducible, it was considered that there was no contamination.

The various targets examined, and whether they were in the form of single crystals, polycrystalline samples, or compressed powders is indicated in Tables 1 and 2. The polycrystalline targets were prepared by arc melting; no binders were used in preparing the compressed powders. To determine if there was any difference in the emission bands from single crystals and compressed powders, the metal M_V and carbon emission bands from the single crystal of NbC were compared with the same bands from an NbC target formed of compressed powders. No difference was observed in the emission bands from the two targets. The determination of x in the MC_x carbides by sample measurements from one position of the target was not adequate. This is because in MC_x carbides the value of x varied throughout the target. For example, the intensity of the C K band from a single crystal of NbC_x (1/4 in. dia by 1/2 in. long) showed that the value of x at one end was 0.95 and 0.85 at the other. The x values were determined by lattice parameter measurements. Pure stoichiometric carbides could not be obtained for groups IV and V transition metal carbides and the highest amount of carbon present in these carbides corresponds to approximately $MC_{0.95}$ which is referred to as stoichiometry in the text.

The samples of Fe_3C examined in the present investigation were in the form of second-phase particles in the Fe-1.83 wt.% C alloy. A micrograph of Fe_3C in the ferrite matrix is shown in Fig. 1a. A "graphitized" Fe target was prepared from a Fe-1.83% C alloy by first

TABLE 1. Carbon K Bands for Stoichiometric Carbides

Material	Peak shift rel. to graphite Δ, eV	Peak wave-length, Å	Half-width $W_{1/2}$, eV	Asymmetry	Formation enthalpy* ΔH°_{298}, kcal/mole	Melting point,* °C	Structure†
Graphite	—	44.85	6	0.83			Hexagonal SC
Diamond	2.1	44.52	8.1	1.25			Cubic
Group A							
TiC	1.95		3.0	1.1	—43	3200	NaCl type SC
VC	1.8		3.3	1.45	—28	2850	NaCl type CP
Cr₃C₂	1.9		3.3	1.6	—21	1870	Orthorhombic CP
ZrC	2.05		2.4	0.85	—44	3530	NaCl type PC
NbC	1.9		2.4	1.05	—33	3500	NaCl type SC
HfC	2.0		3.0	0.80		3880	NaCl type CP
TaC	1.2		3.0	0.80	—78	3880	NaCl type CP
Group B							
(MoCo)₄C	1.2		5.2	0.9			Cubic PC
Fe₃C	1.8		4.4	1.25	+5.98	1650	Orthorhombic‡
Mo₂C	1.2		4.4	1.2	+4.2	2600	Hexagonal CP

*Taken mostly from P. T. B. Shaffer, High Temperature Materials, No. 1: Materials Index, IFI/Plenum, New York (1964); L. S. Darken and R. W. Curry, Physical Chemistry of Metals, McGraw-Hill, New York (1953), p. 364.

†SC denotes a single crystal, PC a polycrystalline sample, and CP a compressed powder.

‡Present as second phase in iron–carbon alloys.

heating to 980°C for 5 min and then quenching. It was then reheated to 700°C and held there for 148 h after which time only 20% of the carbon had been "graphitized." It was then cold-rolled to a 65% reduction in thickness and reheated for another 148 h. Microscopic examination showed that this procedure served to convert all of the carbide to carbon + ferrite. The micrograph of the "graphitized" Fe target is shown in Fig. 1b. The graphite target consisted of spectroscopic electrode graphite with an impurity concentration of 6 ppm.

Experimental Results

Carbon K Bands of Graphite and Diamond

Spectrometer data showing the C K emission bands from diamond, graphite, TiC, VC, and Fe₃C are presented in Fig. 2. The peak heights of the C K bands of diamond and the carbides have been normalized to the peak height of graphite. The diamond was coated with aluminum to prevent charging due to the electron beam. It may be seen that the C K band of diamond

TABLE 2. Characteristics of Ti and V L_{III} Bands

Material	$W_{1/2}$, eV	Asymmetry	Peak shift Δ, eV	L_{II}/L_{III}	Low-energy peak	Structure‡
Ti	4.5	1.3	—	0.25	No	P.C.
TiC₀.₈₃	5.0	1.0	—0.6*	0.15	Yes	P.C.
TiC	5.0	1.0	—1.2*	0.15	Yes	S.C.
V	5.5	1.3	—	0.22	No	P.C.
VC	6.5	1.0	—1.0†	0.17	Yes	C.P.

*Shift relative to Ti L_{III} band of pure Ti.

†Shift relative to V L_{III} band of pure V.

‡See Table 1 for explanation.

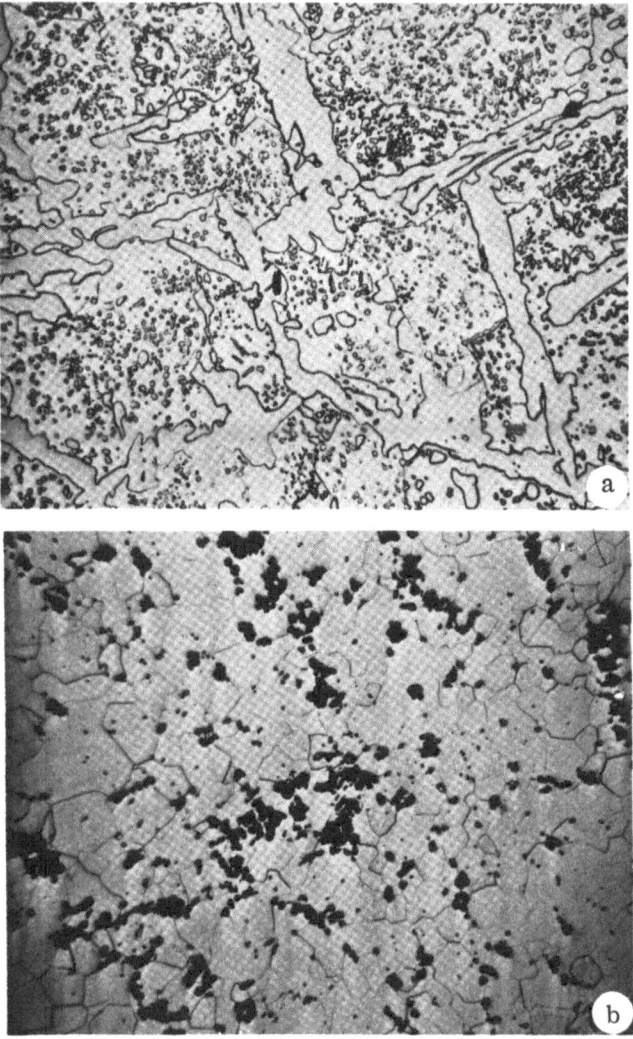

Fig. 1. a) Fe_3C + ferrite formed from 1.83 wt. % C⁻Fe alloy (500×); b) graphitized Fe formed from 1.83 wt. % C⁻Fe alloy (500×).

does not have any structure on the high-energy side of the main peak but does have structure on the low-energy side which is just opposite to the spectrum from graphite. In the case of graphite, there are two distinct humps on the short-wavelength side of the main peak c at 44.85 Å; the most pronounced a is at approximately 44 Å with a weaker hump b at 44.5 Å. There is also a peak f on the long-wavelength side of the main peak at approximately 45.9 Å. For the diamond C K band the most pronounced secondary peak is e at approximately 45.8 Å with the other much weaker hump d at 45.3 Å.

In addition to these differences in structure, it may be noted that the diamond C K peak has shifted 0.34 Å (2.1 eV) toward shorter wavelength relative to that of the peak of the C K band from graphite. The peak wavelengths, $W_{1/2}$ and the index of asymmetry for the C K bands of diamond and graphite are shown in Table 1. The index of asymmetry is defined as the ratio of the part of full width at half maximum lying to the long-wavelength side of the maximum ordinate to that on the short-wavelength side [10].

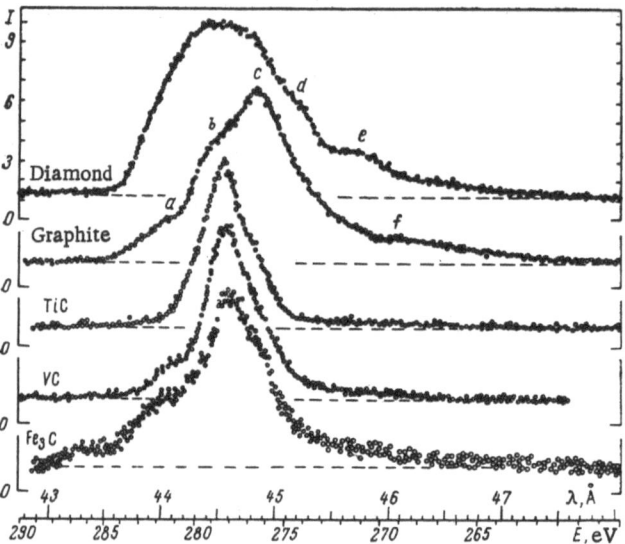

Fig. 2. Carbon K emission band, 2p →1s transition (peaks normalized), for diamond, graphite, TiC, VC, and Fe_3C for a target potential of 4000 V, a beam current of 1.4 mA, a deviation of ±1% for graphite, diamond, TiC, VC, and a deviation of ±3% for Fe_3C. I is the relative intensity.

Although the above results are in general agreement with early investigators [5–9] and the recent work of Sagawa [11] who used a CuBe photomultiplier, some differences do exist which require comment. For example, the ratio of the main peak c to hump a is 0.3 for the C K band of graphite in Fig. 2 whereas Chalkin and Sagawa reported a ratio of 0.5 and 0.6, respectively. It is interesting to note that a ratio of 0.45 was observed in this work for the C K band from carbon deposited by the electron beam. The C K emission band from carbon deposited by the electron beam is shown in Fig. 3. It will be seen that it has the same general shape and peaks a, b, c are at the same wavelength as the C K band from a graphite in Fig. 2. However as indicated above, peak a has a higher intensity relative to the main peak c than for the graphite case. Peak f is not as pronounced for the C K band from deposited carbon. Also its $W_{1/2}$ is larger than for graphite. It is known from electron diffraction that the deposited carbon is amorphous and is probably not pure carbon. The $W_{1/2}$ of the graphite C K bands shown in Table 1 is narrower than that reported by Renninger [5], Skinner [8], and Chalklin [9]. They

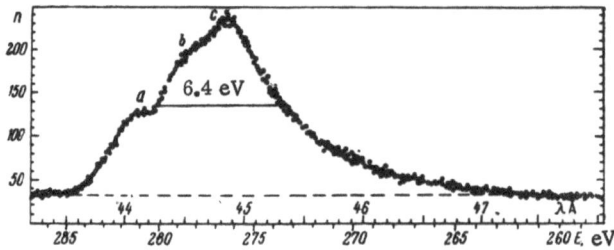

Fig. 3. Carbon K emission band for the carbon deposited by action of electron beam for a target potential of 4000 V, a beam current of 1.4 mA, and a deviation of ±1% (n is the number of pulses per second).

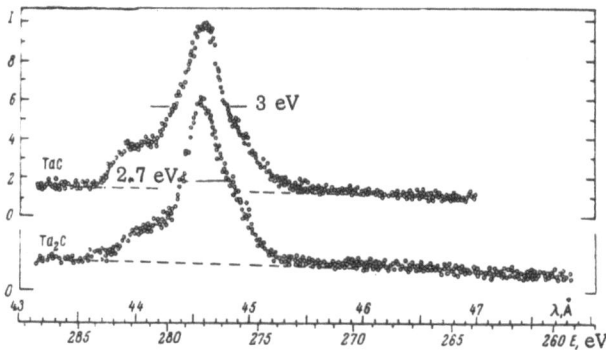

Fig. 4. Carbon K emission band, 2p →1s transition (peaks normalized), for TaC and Ta_2C. The target potential was 4000 V, and the beam current was 1.4 mA. TaC is a NaCl-type structure and Ta_2C is hexagonal. The standard deviation was ±1.5%.

also reported that the $W_{1/2}$ was smaller for the diamond C K band than for the graphite C K band. Just the opposite is the case for the diamond and graphite bands indicated in Table 1. The $W_{1/2}$ of the graphite C K band is narrower than that of diamond and the C K band of the above investigation, because hump a at 44 Å for the graphite C K band in Fig. 2 is below the half maximum intensity. If hump a were above the midpoint, the $W_{1/2}$ of graphite C K band would be greater than the $W_{1/2}$ of the diamond C K band. Renninger [5] and Chalklin [9] show less of a shift of the diamond C K band relative to that of the graphite C K band than indicated in Table 1. Renninger reported a shift of 0.37 Å and Chalklin 0.2 A.

Recently several investigators [12–14] have measured the C K bands of diamond and graphite with a lead stearate crystal in place of a grating. Henke [12] has reported a C K band with a shape close to that measured with a grating. However, the band is less resolved and hump b at 44.5 Å is not observable. Fischer and Baun [13] using a lead stearate analyzer re-

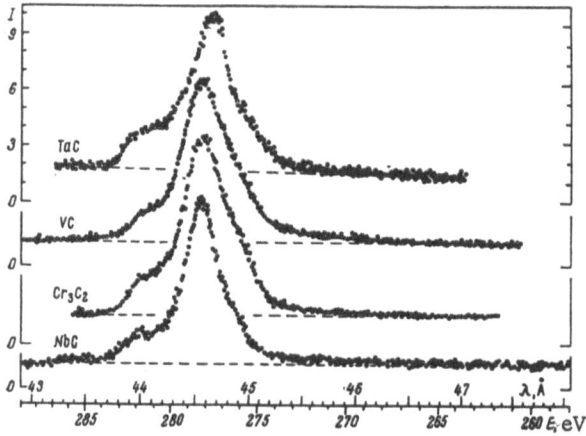

Fig. 5. Comparison of carbon K emission (peaks normalized) bands for group V transition metal carbides TaC, VC, and NbC. Note the similarity in the shape of the peak. The C K band from Cr_3C_2 has been included with group V carbides because it has a shape similar to group V carbides.

ported a shape of the C K band from graphite that is not in agreement with Henke [12] or those
obtained with a grating. Since their wavelengths are not in agreement with ours and Chalklin
[9], it is hard to compare peaks. Their C K band shows only two peaks with the first peak
having the highest intensity. Difference in target cannot be the answer since their actual targets were examined with our spectrometer and the shape of the C K band was the same as
that indicated in Fig. 2. Since the basic difference between Henke's and Fischer and Baun's
spectrometers is that Henke uses x-ray excitation whereas Fischer and Baun employ electron
excitation, it is possible that Fischer and Baun have a contamination problem that is particularly
severe with graphite.

Carbon K Band of Transition Metal Carbides

The C K bands of the carbides of Ti, V, Fe are shown in Fig. 2 and for the carbides of
Ta in Fig. 4. The data on peak position half-width and asymmetry index obtained from these
measurements are included in Table 1. The C K band in Fig. 2 from TiC is a single nearly
symmetrical peak with the possibility of a low-intensity peak on the low-energy side. The previously reported [1, 15] C K band from ZrC is also a single nearly symmetrical peak. There
is no evidence of this peak for the C K band obtained previously at lower resolution with the
2160 grooves/mm grating. The peak wavelength of the C K band from TiC is close to that of
diamond, shifting 0.05 Å (0.3 eV) toward longer wavelengths relative to the diamond C K band.
For the C K band from VC the peak occurs at nearly the same wavelength as the C K band
from TiC and corresponds in wavelength to the hump b on the high-energy side of the main C K
band peak from graphite. The main diamond peak occurs at approximately the same wavelength
as peak b. There is another peak for the C K band from VC that is not present for the C K
band from TiC. This peak occurs at approximately 44 Å and corresponds to hump a for the
C K band from graphite. A similar peak for the C K band from NbC has previously been reported [15] (Fig. 5). The C K band from TaC shown in Fig. 4 also shows this low-intensity
hump. Fischer and Baun [13] also observed this hump for TaC using a lead stearate analyzer
but they showed it having a higher intensity relative to the main peak.

In general, the shapes of the C K bands from the transition metal carbides for a given
group are nearly the same. The shape of the C K bands for group IV transition metal carbides

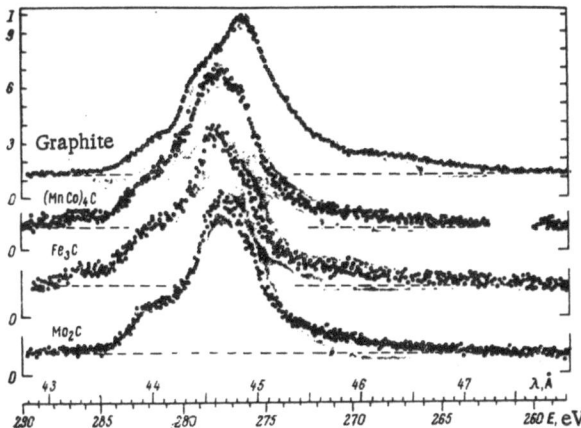

Fig. 6. Carbon K emission bands (peaks normalized) for group VI and higher transition metal carbides compared to the C K band from graphite.
The target potential was 4000 V, the beam current
was 1.4 mA, and the deviation was ±1%.

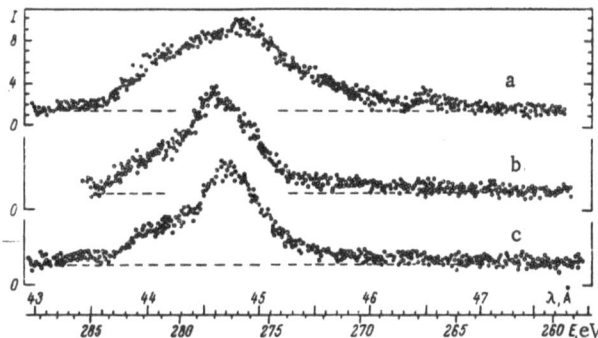

Fig. 7. Carbon K emission bands: (a) C K band for graphitized Fe formed from Fe−1.83% C alloy; (b) C K band for Fe_3C formed from the Fe−1.83% C alloy; (c) C K band, second run on same Fe_3C target.

is a single, nearly symmetrical peak while for group V the C K bands have narrow main peaks with a hump on the higher side. The C K bands for group V carbides TaC, VC, and NbC are compared in Fig. 5. The similarity of the shape of the C K bands of group V carbides is clearly seen from Fig. 5. The C K band from Cr_3C_2 which is a group VI carbide has been included with group V carbides because the shape and peak wavelength is much closer to group V carbides than group VI and higher carbides which are shown in Fig. 6. For group VI and higher the C K bands are broad with a shape close to that of graphite. This is seen for the C K band from $(MnCo)_4C$, Fe_3C, and Mo_2C in Fig. 6. The shapes of these carbide C K bands are closer to that of the graphite C K band than the C K bands from group IV and V carbides. The $W_{1/2}$ of the C K bands in Table 1 shows that the C K bands from group VI and higher carbides are all broader than the C K bands from group IV and V carbides. Thus, as the group number increases for a given transition metal period, the C K band changes from a narrow nearly symmetrical peak to a broad band with a shape approaching that of graphite. This can be seen for the first series transition metal carbides in Fig. 2. The fact that the $W_{1/2}$ of the carbides is narrower than for graphite was first pointed out by Renninger [5] as a result of his work on the C K band from Cr_3C_2.

As could be expected from a comparison of the C K bands of graphite and Fe_3C, a variation in the $W_{1/2}$ and wavelength was found between the C K band from a "graphitized" Fe and Fe_3C. The C K band from graphitized Fe is shown in Fig. 7a and it may be seen that the shape of the C K band from "graphitized" iron is similar to that of the C K band from graphite shown in Fig. 2 but is not identical. Both curves have the same peak wavelength and humps, a and b, on the high-energy side of the main peak. However, hump a has an intensity of 0.65 that of the main peak for "graphitized" Fe compared to a ratio of 0.3 for the C K band in Fig. 2. Because hump a has a greater relative intensity for the "graphitized" case, the $W_{1/2}$ for the C K band is broader (8.3 eV) from graphitized Fe than for the graphite target (6 eV). These results suggest that there is some difference between electrode graphite and carbon precipitation in iron. In order to study the C K band from carbon in solid solution the Fe−1.83% C alloy was heated to 1200°C and quenched. The micrograph of this target before electron bombardment is shown in Fig. 8a and contains approximately 25% martensite in austenite. However, by observing the shift in the peak wavelength of the C K band it was found that the carbon had precipitated as Fe_3C after approximately 5 min of electron bombardment. Comparison of the shape and wavelength of the C K band from this target shown in Fig. 7b with that from Fe_3C in Fig. 2 shows that the spectra are the same. Further evidence that carbon has precipitated as Fe_3C is shown by the micrograph in Fig. 8b. X-ray diffraction measurements also showed that carbon was

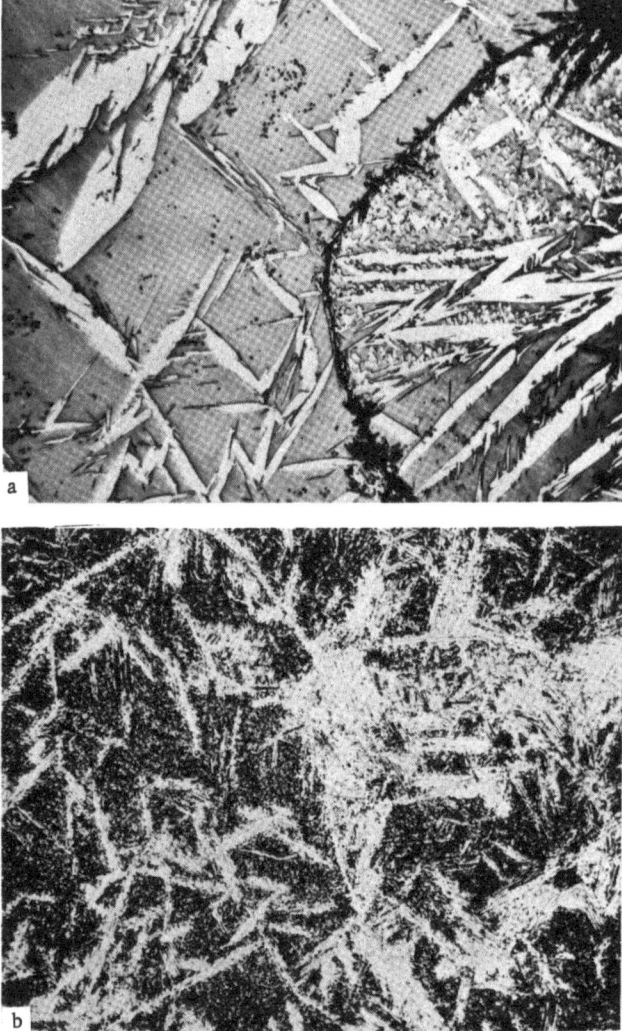

Fig. 8. Micrograph of Fe–1.83% C alloy quenched
from the completely austenitic condition: a) the
top micrograph shows the target before electron
bombardment; b) bottom micrograph shows the target after approximately 5-min electron bombardment.

precipitated as Fe$_3$C. The C K band in Fig. 7c was obtained after approximately 12 h of electron bombardment of this target. The peak wavelength of the bottom curve in Fig. 7c has shifted
0.8 eV toward the peak wavelength of graphite and the main peak is more symmetrical than is
the curve in 7b. These changes in the curve in Fig. 7c could reflect the fact that some of the
Fe$_3$C is transforming back to austenite due to electron beam heating. The above results show
that changes in shape and wavelength can be used to study transformations in the Fe–C system.

Metal Emission Bands of Carbides

The Ti L$_{II, III}$ emission bands (3d + 4s → 2p transition) from Ti, TiC [1], TiC$_{0.83}$ are
shown in Fig. 9 and the V L$_{III}$ band from V and VC in Fig. 10. Comparison of the shape of the
L$_{II, III}$ emission band for Ti and V indicates that their shapes are very much alike. Similarly
the shapes of the L$_{II, III}$ emission bands for TiC and VC are also nearly the same, e.g., the

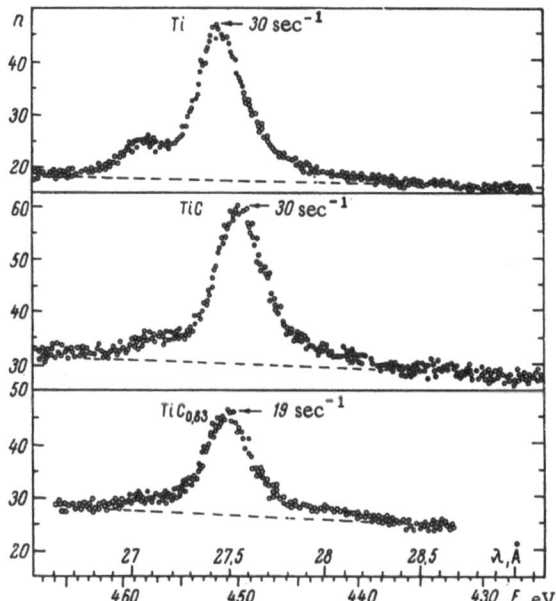

Fig. 9. Titanium $L_{II, III}$ emission bands ($3d +$ $4s \rightarrow 2p$ transition) for Ti, TiC, and $TiC_{0.83}$. The target potential was 4000 V and the beam current was 1.4 mA. The standard deviation was $\pm 1.5\%$. Note shifts in the L_{III} peaks. The ordinate gives n in pulses/sec.

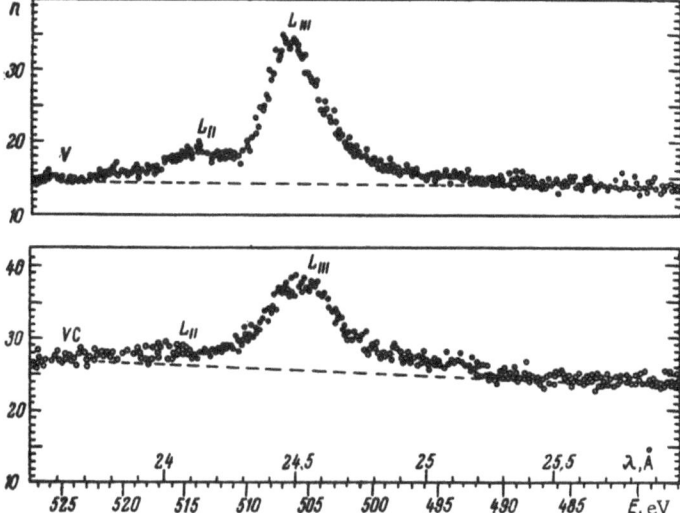

Fig. 10. Vanadium $L_{II, III}$ emission band ($4s + 3d \rightarrow$ $2p$ transition) for V and VC. The target potential was 4000 V and beam current was 1.4 mA. Note the shifts in the L_{III} peaks. The ordinate gives n in pulses/sec.

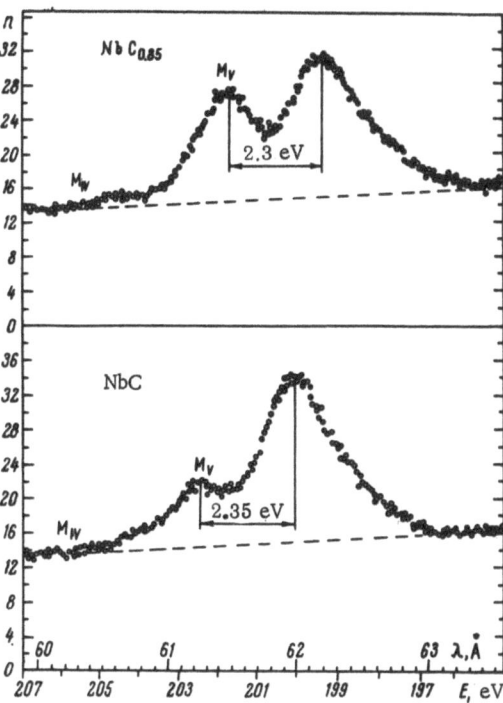

Fig. 11. M_{IV} and M_V emission band (5p → 3d transition) for single crystals of NbC and NbC$_{0.85}$. The target potential was 4000 V and the beam current was 1.4 mA. The deviation was ±1.5%. The ordinate gives n in pulses/sec.

low-intensity peak on the low-energy side of the L$_{III}$ band is present for both carbides. However, from the characteristics of the bands as summarized in Table 2, the L$_{III}$ emission band for vanadium is broader and has a lower index of asymmetry than the Ti L$_{III}$ band from titanium. These differences could be due to the increase in the broadening effect of the 2p level and instrumental error [3].

In comparing the L$_{II, III}$ emission band from the metal to that from its carbide, it was found that W$_{1/2}$ for the V L$_{III}$ band is 25% larger in vanadium compared to VC in contrast to only a 10% difference between Ti and TiC. These differences in W$_{1/2}$ could be related to the similar differences in the W$_{1/2}$ of the C K bands between TiC and VC. It will be noted that the L$_{III}$ bands of the metals have asymmetrical peaks while their carbides all have symmetrical peaks. The occurrence of asymmetrical peaks with an emission edge from metals and symmetrical peaks and no emission edge from their carbides is also characteristic of the M$_V$ bands (5p → 3d transition) for ZrC, NbC, and Mo$_2$C [15]. Lukirski et al. [17] recently reported symmetrical peaks for the M$_V$ emission band from NbC.

The effect of departure from stoichiometry on the M$_V$ emission bands from NbC and NbC$_{0.85}$ is shown in Fig. 11. It will be noted that contrary to the case of TiC$_x$ carbides, here the metal emission band has changed considerably for the nonstoichiometric condition. In particular, in NbC$_{0.85}$ the peaks of the M$_V$ bands have nearly the same height and are shifted to lower energies relative to the M$_V$ band from NbC. This indicates that the M$_V$ emission bands in NbC carbides are more sensitive to defect structures than the Ti L$_{II, III}$ bands in the TiC$_x$ case. It will also be observed from Table 3 and Fig. 9 that the shape of the C K band from TiC$_{0.83}$ is the same as that for TiC, although there was an increase in the index of asymmetry for the C K band from $TiC_{0.83}$ relative to the C K band from TiC.

Discussion

Bonding

Table 1 lists the characteristics of band shape and wavelength measured on the present spectrometer as well as the melting temperatures and the heats of formation. Because of statistical deviation in the x-ray data the error in determining the peak wavelength and $W_{1/2}$ is ± 0.05 eV. The transition metal carbides have been divided into two groups according to how strong the bond is. Group A carbides are the strongly bonded carbides with high negative heats of formation whereas group B are weakly bonded carbides with positive heats of formation. The carbides in group A all have relatively narrow $W_{1/2}$ compared to that of graphite. In an earlier work Broili et al. [6] indicated that there was a relation between the width of the C K band from carbides and bonding. With the exception of TaC the peak wavelength is close to that from the C K band of diamond. For the group B carbides the $W_{1/2}$ is larger and the peak wavelength has shifted closer to the peak wavelength of graphite. In general, the closer the shape of the carbide C K band is to that of graphite the weaker the bonding. With the exception of Cr_3C_2, group A carbides consist of group IV and V carbides while group B carbides consist of group VI and beyond. Further justification for the relation between C K band shape and bonding is shown by the case of Cr_3C_2. From the bonding and other properties [16] it has been found that Cr_3C_2 should be classed with group V carbides rather than with group VI carbides. As shown above the shape and wavelength of the C K band from Cr_3C_2 fits with group V carbides rather than with group VI and higher carbides.

Another interesting correlation between bonding and band shape can be observed from Table 1. Within a given period for the group A carbides the index of asymmetry increases as the bonding becomes weaker. A similar situation is found for the MC_x carbides shown in Table 3 where it will be noted that as x decreases for a given M there is an increase in asymmetry and a decrease in the bond strength. From these $W_{1/2}$ measurements it appears that for a decrease in x, $W_{1/2}$ remains constant unless there is a change of structure. No definite conclusions can be drawn from the wavelength shift measurements in Table 3.

Another important area of correlation is the relation between crystal structure, band shape, and bonding. In Table 1 all of group A carbides with the exception of Cr_3C_2 have the NaCl-type structure. However, as shown in Table 3, $NbC_{0.5}$ and $TaC_{0.5}$ have sufficiently high

TABLE 3. C K Bands for MC_x Carbides

Carbide	Peak shift* Δ, eV	Half-width $W_{1/2}$, eV	Asymmetry	Melting point, °C	Structure †
$TiC_{0.95}$	1.85	3.0	1.1	3200	NaCl type SC
$TiC_{0.83}$	1.9	3.0	1.45	3100	NaCl type PC
$NbC_{0.95}$	1.9	2.4	1.05	3500	NaCl type SC
$NbC_{0.85}$	1.7	2.5	1.3	3200	NaCl type PC
$NbC_{0.5}$	1.6	3.0	1.2	3090	Hexagonal CP
$TaC_{0.95}$	1.2	3.1	0.80	3880	NaCl type SP
$TaC_{0.5}$	1.7	2.7	1.7	3400	Hexagonal CP

* Peak shift relative to the C K band of graphite.
† SC denotes a single crystal, PC a polycrystalline sample, and CP a compressed powder.

bond strengths to be classed with group A carbides yet they both have hexagonal structures. Thus, it is not possible to associate high bond strength with a given crystal structure. Neither does it appear possible to make broad generalizing comparisons between crystal structure and band shape. It has been previously observed [1] that C K emission bands from transition metal carbides with the NaCl-type structure generally have narrow nearly symmetrical peaks while emission bands from hexagonal structures are broader and show a greater index of asymmetry. The C K band from hexagonal $TaC_{0.5}$ shown in Fig. 4 is more asymmetrical than from NaCl-type TaC but it has a narrower half-width. In addition, there is a large difference in shape between the C K band from $TaC_{6.5}$ (Fig. 4) and the C K band from hexagonal Mo_2C (Fig. 6). Although the local arrangement of atoms has an influence on the shape of the band, it will be seen from Table 1 that bonding has a much greater influence on the shape of the band.

Electron Distribution

It has been shown [1] that, as a result of the shift of the Ti L_{III} band toward higher energy for TiO_x compound with increasing x, the shift of the peak of the Ti L_{III} band toward lower energy (shown in Table 4) for TiC_x with increasing x is due to the Ti atom becoming more negative. The fact that the titanium atom is possibly more negative in TiC than in titanium metal is based on the principle that the Ti L_{III} shifts toward higher energy when the titanium atom is becoming more positive as a result of electron transfer to oxygen in TiO_x compounds. Thus, a shift of the Ti L_{III} band toward lower energy indicates that the Ti atom is becoming more negative. Faessler [18] has had considerable success in relating wavelength shift with electron distribution. It will be noted from Table 4 that the V L_{III} band also shifts toward lower energy for VC and higher energy for V_2O_3 relative to vanadium. By a reasoning process similar to that used for the titanium atom it would appear that the vanadium atom is also more negative in VC than in pure vanadium metal. The idea that charge is transferred from carbon to the metal atom in transition metal carbides has been advanced by Kiessling [19], Robbins [20], Dempsey [21], Costa and Conte [22], and Williams and Lye [23]. Their conclusions are based on both theoretical and experimental consideration. A transfer of electrons from the metal to carbon atom would indicate a "semi-ionic" character to the carbon metal bond, which is discussed by Williams and Lye [23].

If carbon is transferring electrons to titanium and vanadium atoms then the transfer should be reflected in the wavelength of the C K bands from TiC and VC. A comparison of the peak wavelength of the C K band from TiC and VC with the C K band from a natural carbon atom in diamond (Table 1) shows that their peak wavelengths are close to that of diamond. If carbon is transferring electrons to the metal it would be expected that the peak wavelengths of the C K band from TiC and VC would be closer to that of graphite whose carbon atom has three bonding electrons and approximately one electron for conduction. However, changes in electron

TABLE 4. Shifts in Peaks of Ti and V L_{III} Bands

Material	ΔE, eV	Peak shift L_{III} $\Delta\lambda$, Å	Material	ΔE, eV	Peak Shift L_{III} $\Delta\lambda$, Å
Ti L_{III} (a)•			Ti_2O_3	+2.3	−0.14
			$TiO_{1.97}$	+3.5	−0.20
$TiC_{0.95}$	−1.2	+0.07			
$TiC_{0.93}$	−1.0	+0.065	V L_{III} (b)•		
$TiC_{0.83}$	−0.6	+0.04			
Ti	—	—	VC	−1.0	+0.05
$TiO_{0.9}$	+0.9	−0.06	V	—	—
$TiO_{1.17}$	+1.8	−0.11	V_2O_3	+1.9	−0.09

• a) Ti L_{III} peak shift relative to Ti L_{III} band of pure Ti; b) V L_{III} peak shift relative to V L_{III} band of pure V.

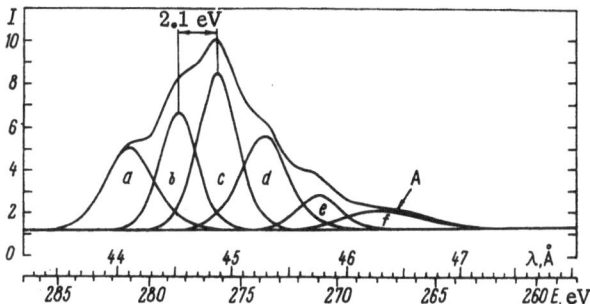

Fig. 12. The C K emission band from Fig. 3 resolved into five Gaussian peaks and one non-Gaussian peak. Peaks a, b, c, d, e, and f correspond in wavelength to humps a, b, c, d, e, and f of the C K band for graphite and diamond in Fig. 2. Peaks d, e, and f are more pronounced than for the C K in Fig. 3 in order to show their relation to these peaks on diamond and graphite C K bands in Fig. 2. Curve A is the non-Gaussian distribution.

distribution within the carbon band may be such that changes in peak wavelength cannot be used as an indication of the degree of ionization. This is shown in Fig. 12 where the carbon band in Fig. 3 has been resolved into five Gaussian and one non-Gaussian peaks by a Dupont curve analyzer. These peaks correspond in wavelength to humps a, b, c, d, e, and f shown for the diamond and graphite C K bands in Fig. 2. Peaks d, e, and f are more pronounced in Fig. 12 than for the C K band in Fig. 3, in order to show their correspondence with humps d, e, and f in diamond and graphite C K bands in Fig. 2. The shape of the C K bands in Figs. 2 and 5 can be duplicated by assigning the proper weights to peaks a, b, c, d, e, and f. For example, the C K band from TiC can be duplicated by giving a strong weight to Gaussian peak b and a very small weight to peak c to account for the asymmetry in the C K band of TiC. The rest of the peaks have zero or nearly zero weight. In the case of group V carbides in Fig. 5 Gaussian peak b is still predominant but Gaussian peak a is now in evidence. The hump on the low-energy side of the Cr_3C_2 C K band can be accounted for by an increase in the weight given to Gaussian peak c. For the diamond C K band Gaussian peak b has the strongest weight and peak c has the strongest weight for the graphite C K bands. It is of interest to note that the energy difference between peaks b and c in Fig. 12 is the same as the energy difference between the peaks of the diamond and graphite C K bands. Thus, the wavelength shift appears to be due to changes in weighting of the peaks in Fig. 12 and not a shift of the 2p level.

The Gaussian peaks in Fig. 12 correspond closely with the sub-bands reported by Sagawa [11]. Gaussian peak a corresponds to the π sub-band; peak b to the σ_3 sub-band; peak c to the σ_2 sub-band; peaks d and e to the σ_1 sub-band; and non-Gaussian peak f to the Sagawa satellite. It must be emphasized that the above picture is only qualitative.

Conclusions and Summary

The above results suggest that differences in the shape of the C K band from transition metal carbides can be correlated with bonding. However, crystal structure does not appear to be as readily correlated with band shape. Although the correlation of changes in shape of the metal emission bands with changes in bonding is not as clear as it is for the C K bands, it is possible to correlate shifts in the peak of the metal emission with changes in electronic charge on the metal atom. However, it is not possible to correlate directly changes in wavelength of the C K band with changes in charge on the carbon atom due to large changes in electron distribution within the carbon band. It must be emphasized that more work is required on the relation between emission bands, bonding, and electron distribution, especially of a theoretical nature.

Acknowledgments

The author would like to thank A. Panson of Westinghouse Research Laboratories for supplying the TiO_x targets and also Robert G. Lye of the Martin Co. for supplying the TiO_x targets. At this laboratory the author wishes to thank the following members: W. A. Hester for assistance with the experiments; L. Zwell and E. J. Fasiska for the x-ray diffraction work; and R. A. Grange for supplying the Fe−C alloy and for his helpful discussion on the Fe−C system.

Addendum

Since the presentation of this paper additional experimental information on transition metal carbides indicates that the peak shift of the Ti L_{III} band should be interpreted differently than was done in the present paper. Recent experimental results by soft x-ray spectroscopy and ESCA (electron spectroscopy by chemical analysis) indicate that if there is ionic character in TiC, electrons are being transferred from titanium to carbon. A complete discussion of this is in print [24].

Literature Cited

1. J. E. Holliday, in: Advances in X-Ray Analysis (Proc. Fourteenth Conf., University of Denver, 1955), Vol. 9, Plenum Press, New York (1966), p. 365.
2. J. E. Holliday, Rev. Sci. Instrum., 31:891 (1960).
3. J. E. Holliday, "Soft x-ray spectroscopy in the 10–150 Å region," in: Handbook of X-Rays (ed. by E. F. Kaelble), Ch. 38, McGraw-Hill, New York (1967).
4. J. E. Holliday, in: Developments in Applied Spectroscopy (Proc. Sixteenth Mid-America Symposium on Spectroscopy, 1965), Vol. 5, Plenum Press, New York (1966), p. 77.
5. M. Renninger, Z. Phys., 78:510 (1932).
6. H. Broili, R. Glocker, and H. Kiessig, Z. Phys., 92:27 (1934).
7. M. Siegbahn and T. Magnusson, Z. Phys., 96:1 (1935).
8. H. W. B. Skinner, Proc. Roy. Soc., London, A239:95 (1940).
9. F. C. Chalklin, Proc. Roy. Soc., London, A194:42 (1948).
10. J. A. Beardon and C. H. Shaw, Phys. Rev., 48:24 (1935).
11. T. Sagawa, J. Phys. Soc. Japan, 21:49 (1966).
12. B. L. Henke, in: Advances in X-Ray Analysis (Proc. Fourteenth Conf., University of Denver, 1955), Vol. 9, Plenum Press, New York (1966), p. 430.
13. D. W. Fischer and W. L. Baun, in: Advances in X-Ray Analysis (Proc. Fourteenth Conf. University of Denver, 1955), Vol. 9, Plenum Press, New York (1966), p. 329.
14. R. A. Mattson and R. C. Ehlert, in: Advances in X-Ray Analysis (Proc. Fifteenth Conf., University of Denver), Vol. 10, Plenum Press, New York (1967).
15. J. E. Holliday, in: The Electron Microprobe, Wiley, New York (1966), p. 3.
16. H. Nowotny, private communication.
17. A. P. Lukirskii and T. M. Zimkina, Izv. Akad. Nauk SSSR, Ser. Fiz., 27:330 (1963).
18. A. Faessler, in: Proc. Tenth Intern. Colloq. on Spectroscopy, University of Maryland, 1962, publ. by Spartan Books, Washington, D. C. (1963), p. 307.

19. R. Kiessling, Met. Rev., 2:77 (1957).

20. D. A. Robbins, Powder Met., No. 1/2, p. 172 (1958).

21. E. Dempsey, Phil. Mag., 8:285 (1963).

22. P. Costa and R. R. Conte, Met. Soc. AIME Inst. Metals Div. Spec. Rept., AIME, New York, 1964, No. 13, pp. 3-27.

23. W. S. Williams and R. G. Lye, Research to Determine the Mechanisms Controlling the Brittle-Ductile Behavior of Refractory Cubic Carbides, U. S. Air Force Tech. Doc. Rep. ML-TDR-64-25, Part II (March, 1965).

24. J. E. Holliday, J. Phys. Chem. Solids, in press (1971).

INVESTIGATION OF THE DISTRIBUTION
OF THE POTENTIAL IN THE UNIT CELL
OF LITHIUM FLUORIDE*

L. V. Shevtsov and V. P. Tsvetkov

An experimental investigation was made of the x-ray atomic scattering factors of the components of lithium fluoride. These factors were used to plot the distributions of the potential energy in the unit cell of LiF along three principal directions representing the lines joining the following nearest neighbors: Li—Li, Li—F, and F—F. An analysis of these distributions was used to determine the nature of the chemical bonds in the crystal lattice of lithium fluoride.

Havighurst [1] was the first to use the distribution of the electron density in a crystal to determine the nature of the chemical bonds. He used three-dimensional Fourier series to plot the electron density distributions along the cube edges in NaCl, NaF, and LiF. He concluded that ionized atoms were present in these compounds. The experimental results and the method used were refined by Havighurst in his later work [2—4]. Other workers used the radial density distributions (in addition to the three-dimensional distributions) and repeated Havighurst's investigations [5, 6]. In spite of the fact that the numerical differences were only small, the results of all these studies have been interpreted in different ways. The results obtained for LiF were of special interest. It would seem that an investigation of the lighter elements should give more definite results but this was not observed for lithium fluoride. Thus, Havighurst [3] compared the electron density along the Li—F direction with the density along the corresponding direction in NaCl and found that the density $\rho(r)$ in LiF did not vanish at any point along this direction, whereas zero density was found for sodium chloride. Havighurst attributed the absence of zero values of $\rho(r)$ in LiF to the nonsphericity of the component atoms [3].

Krug et al. [5] investigated NaCl and LiF and found a higher than expected value for the electron density along the Li—F direction and they explained this observation by the presence of covalent bonds. However, in a later paper, Krug et al. [6] attributed this observation to the nonsphericity of the atoms and concluded that there were no definite covalent bonds in LiF. These investigations were repeated in 1958 [7] using better methods but no new information was obtained on $\rho(r)$ in LiF, although the new measurements were carried out employing an absolute method.

This indeterminacy may be due to the lack of precision in the measurements of the structure amplitude; to the failure to include all the factors which affect the measurements; or to

---

* "Crystals," pp. 198-207 (see page 3).

the weak sensitivity of the electron density curves to the nature of the chemical bonds. This lack of sensitivity is supported by the following reasoning.

The theoretical values of the atomic scattering factors of ions and neutral atoms differ very little in the range of angles which are usually employed in measurements. This is true in spite of the fact that an ion is assumed to have lost one or more electrons completely. In real solids, the formation of an ion is accompanied by a slight shift of its electrons over distances not exceeding 20% of the atomic radius so that such shifted electrons continue to participate in the scattering of x rays. Any influence of such a shift on the atomic scattering factor is eliminated in the plotting of the electron density because the distribution obtained is determined by the sum of the electron densities of the atoms in a given unit cell. The pattern in all the other cells is a simple repetition of the one being considered. However, if we analyze the potential within the cell and not the electron density, we may expect the changes in the f curves, associated with the charge distribution, to affect considerably the distribution of the potential because at each point the potential represents the sum of the contributions of the potentials of many atoms and, therefore, the effect in question is enhanced.

In an earlier investigation of silicon and copper [10, 11], Gorokhov and one of the present authors demonstrated that the potential curves for various types of bond differed more strongly than the electron density distributions plotted from the same experimental values of F_{hkl}. In view of this, it seemed desirable to repeat the earlier investigations of lithium fluoride to obtain the potential distribution in its lattice.

The intensities of the Bragg reflections were measured using monochromatic Mo K_α radiation and a scintillation counter connected to a coincidence circuit [12]. The primary beam was sufficiently stable so that the counting error in any time interval did not exceed $1/\sqrt{N}$. The whole counting system was linear within 0.7% for all the counting rates used in our measurements. When the scanning rate of the counter was 0.5° per minute and the integrator time constant was 4 sec, the ratio of the integrated intensities of the strong to the weak lines, measured at separate points and recorded by a potentiometer, did not vary by more than 0.2%. This made it possible to ignore the delay time introduced by the counting and recording system.

The component of the wavelength equal to half the wavelength of the Mo K_α radiation was eliminated from the primary spectrum by applying a voltage of 33 kV and a current of up to 10 mA to the x-ray tube.

A crystal of pentaerythritol, grown from a solution, was used as a monochromator. The ratio of the (220) and (200) reflection intensities of the LiF powder was determined with and without the monochromator. The results showed that the monochromator crystal produced reflections in agreement with the kinematic theory.

The use of an organic monochromator made the reflected beam practically nondivergent in the horizontal plane. The vertical divergence was 2° when two 4-mm-high entry slits were used. The focusing arrangement was such that the counter slit, the center of the surface of the sample, and the center of the surface of the monochromator crystal (acting as the source of rays) were located on the same circle.

Since lithium chloride is a compound of two light elements, we ignored the extinction in powders whose grain size was $2-3\,\mu$. This grain size was achieved by grinding in a mortar; it was measured with a microscope and also deduced from the x-ray diffraction ions obtained in an RKD camera using unfocused radiation.

The (200) reflection was measured using samples compacted into pellets at pressures of $5-30$ kgf/cm^2. These measurements indicated that the pellets prepared from dry powders at pressures exceeding 10 kgf/cm^2 had a preferential orientation. The addition of a binder to the powder retarded the formation of a texture because it appeared only at pressures exceeding 20 kgf/cm^2.

The preferential orientation (texture) could be destroyed by cutting off a surface layer.

In our experiments, the powder was compacted in a ring-shaped mold, which was used to orient the sample. Therefore, we were unable to cut off the surface layer. The formation of a texture was prevented by adding a very small amount of vacuum grease and by using compacting pressures not exceeding 2–3 kgf/cm^2.

Lithium fluoride had the advantage that all its fluorescent radiation was absorbed completely in air. The background in the diffraction patterns was fairly smooth and practically flat at distances of ±2° from a peak.

The structure amplitudes were calculated from the formula

$$F_{hkl} = \left[\frac{I \mu N_0^2 F_0^2 (HPLG)_0}{I_0 \mu_0 N^2 HPLG\, e^{-2M}} \right]^{\frac{1}{2}},\tag{1}$$

where I is the integrated intensity; μ is the linear absorption coefficient; N is the number of unit cells per unit volume of the substance; e^{-2M} is the temperature factor; H, P, L, and G are, respectively, the multiplicity, the polarization, the Lorentz, and the shape factors; the subscript 0 denotes the standard sample.

Formula (1) is valid if the primary beam is absorbed completely in a sample. This condition was observed when the thickness of the LiF sample was 7 mm. The absorption coefficient μ in Eq. (1) should represent a continuous (nondisperse) sample. The correction for the thermal motion was introduced in the form of the factor

$$\exp\left[-2B \left(\frac{\sin \vartheta}{\lambda} \right)^2 \right],$$

where

$$B = \frac{6h^2}{m_a k \Theta} \left[\frac{\Phi(x)}{x} + \frac{1}{4} \right]; \quad \Theta = 735\,°K\ [13].$$

The structure amplitudes were converted to absolute units by comparison with the (220) reflection of NaCl, whose intensity was measured earlier by an absolute method [7, 8].

The experimental values of the structure amplitudes were in good agreement with the theory of ionic crystals (Fig. 1).

The structure amplitudes were used to calculate the atomic scattering factors of the lithium and the chlorine ions. The experimental values obtained in this way were approximated by the following expressions:

$$f(s) = \frac{2}{1 + 0.025 \left[4\pi \left(\dfrac{\sin \vartheta}{\lambda} \right) \right]^2}\tag{2}$$

for the Li$^+$ ions, and by

$$f(s) = \frac{3}{1 + 0.01 \left[4\pi \left(\dfrac{\sin \vartheta}{\lambda} \right) \right]^2} + \frac{7}{\left[1 + 0.055 \left\{ 4\pi \left(\dfrac{\sin \vartheta}{\lambda} \right) \right\}^2 \right]^2}\tag{3}$$

for the F$^-$ ions.

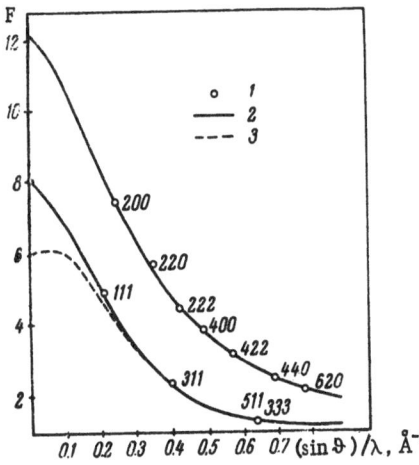

Fig. 1. Structure amplitudes of lithium fluoride: 1) experimental values; 2), 3) theoretical values calculated by the Hartree–Fock method for the Li^+ and F^+ ions and for neutral atoms.

The approximating curves are shown in Fig. 2. The points in that figure represent the experimental values. The approximation of the f curves by expressions such as Eqs. (2) and (3) indicated that the values of f for s = 0 represented ionized atoms although this did not follow directly from the x-ray scattering experiments. The maximum corresponding to $F_{111} = f_F - f_{Li}$ was too far from zero for satisfactory extrapolation of the f curve. Therefore, the f curves near the zero values of $(\sin\vartheta)/\lambda$ could be found only if additional information were available which would make it possible to draw unambiguous conclusions about the correctness of the approximation used. In the case of ionic crystals of the diamagnetic type, this can be done most conveniently using the experimental values of the magnetic susceptibility. The molar diamagnetic susceptibility is given by the following formula

$$\chi_d = -\frac{Ne^2}{6mc^2} \sum \overline{r_i^2}.$$ (4)

The average quantum-mechanical value of the square of the radius $\overline{r_i^2}$ can be calculated if we know the radial distribution of the electrons

$$U(r) = \frac{2r}{\pi} \int_0^\infty sf(s)\sin srds,$$ (5)

where $f(s)$ is the analytic expression for the atomic scattering factor. The values of $f(s)$ can be found by approximating the atomic scattering factors on the assumption that atoms are neutral or ionized. Then, using Eqs. (4) and (5), we can calculate χ_d for a particular approximating curve.

Fig. 2. Curves approximating the atomic scattering factors of the F^- (1) and Li^+ (2) ions.

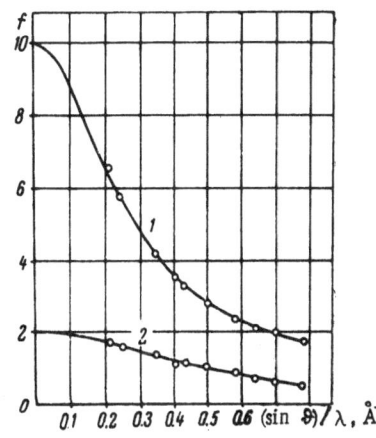

The values of the diamagnetic susceptibility χ_d of various semiconductors were determined by Sirota and his colleagues [21–24]. They used the experimental values of the electron density, from which they deduced values for the susceptibilities χ_d which were comparable with the experimental values.

We carried out the converse operation. The experimental values of the diamagnetic susceptibility were compared with the calculated values obtained using various approximations. It was found that the measured values of χ_d differed severalfold from the values calculated on the assumption that neutral atoms were present in lithium fluoride crystals. However, when we assumed that these crystals consisted of ions, we found that the calculated and the experimental values were practically identical. This indicated that our approximations of the f curves for ions were correct.

The following values of the diamagnetic susceptibility χ_d were obtained by different authors for LiF:

1) the experimental value reported by Dorfman [14] was $10.4 \cdot 10^{-6}$;
2) the susceptibility calculated on the assumption that the fluoride consisted of ions was $10.05 \cdot 10^{-6}$;
3) the assumption that neutral atoms were present gave a calculated value of $36.85 \cdot 10^{-6}$;
4) the value calculated by the Kirkwood method [15] was $10.4 \cdot 10^{-6}$;
5) the calculations carried out using the Slater approach [16] yielded $9.0 \cdot 10^{-6}$.

This approach provided a more convincing proof of the presence of ions in LiF than a simple comparison of the first difference maximum with the theoretical curves for ions and neutral atoms. We were thus able to use Eqs. (2) and (3) with the appropriate parameters in plotting the potential distributions along the [100] and [110] directions in LiF.

The electrostatic potential was calculated from the expression

$$\Phi(R) = \frac{e}{a} \sum_{i=1}^{4} \frac{1}{A_i} - \frac{e}{a} \sum_{j=5}^{8} \frac{1}{A_j} +$$

$$+ \frac{Z_{Li}e}{a} \sum_{i=1}^{4} \frac{e^{-\frac{aA_i}{\alpha_{Li}}}}{A_i} + \frac{Z_{1F}e}{a} \sum_{j=5}^{8} \frac{e^{-\frac{aA_j}{\alpha_{1F}}}}{A_j} + \frac{Z_{2F}e}{a} \sum_{j=5}^{8} \frac{\left(1 + \frac{aA_j}{2\alpha_{2F}}\right)e^{-\frac{aA_j}{\alpha_{2F}}}}{A_j}, \quad (6)$$

where a is the lattice constant; e is the electronic charge; α and Z are the fitting parameters; A_i and A_j are the functions of the coordinates of the atoms at the point at which the potential is to be determined.

The first two terms of Eq. (6) were calculated on a Minsk–14 computer using a program in which seven coordination spheres were taken into account.

The functions $\Phi(R)$ along the Li–Li, Li–F, and F–F directions are plotted in Fig. 3 on two different scales.

It is evident from this figure that at no point did the potential distribution fall to zero or change its sign. Similar results were reported earlier for NaCl [17]. However, in the case of rocksalt, the sign of the potential did not change because the experimental values were approximated by the f curves for neutral atoms. An investigation of KCl showed that the potential energy distribution in the case of K^+ and Cl^- ions had positive regions, which were expected for an ionic lattice with regions of excess negative charge.

The absence of such regions from the LiF lattice could be explained using a feature observed in the electron density distributions $\rho(r)$. The minimum of $\rho(r)$ found along a direction

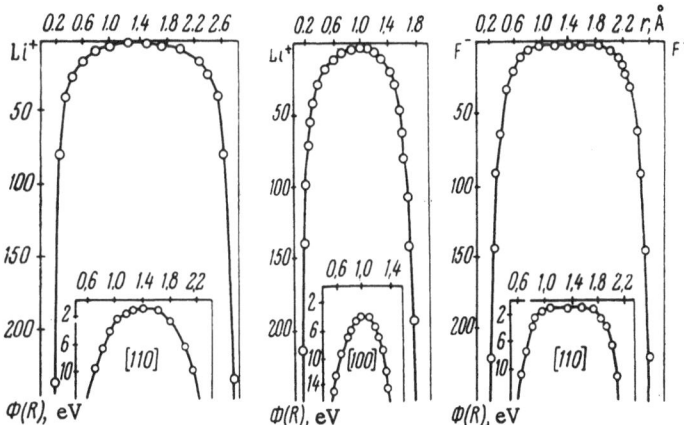

Fig. 3. Potential distributions in the LiF lattice along
the principal crystallographic directions.

joining two neighboring atoms was located at a position which disagreed with the crystallochemical ionic radii. This was first pointed out for LiF in [5] and then for NaCl [18] and KCl
[9].

The following table lists the cation radii in LiF, NaCl, and KCl, the distances of the
$\rho(r)$ minima from the centers of the cations, and the ionic radii r_0 which we deduced from the
radial electron density distributions obtained by approximating the atomic scattering factors
with smooth curves:

Ion	Pauling radius	Goldschmidt radius	Distance to minimum	r_0
Li+	0.60	0.78	0.97	0.70
Na+	0.95	0.98	1.17	0.92
K+	1.33	1.33	1.57	1.60

The method used to calculate r_0 was similar to that described earlier [19, 20]. The experimental values of the atomic scattering factors were approximated by f(s) curves calculated
on the assumption that the atoms were neutral. Having found f(s), we determined the radial
distribution functions in accordance with Eq. (5). These functions were employed to find the
number of electrons outside a sphere of radius r:

$$n(r) = Z - \int_0^r U(r)\,dr = \sum_{i=1}^{2} Z_i \left(1 + \frac{r}{\alpha_i} \right) \exp\left(-\frac{r}{\alpha_i} \right),$$

where Z is the total number of electrons belonging to a given atom; α_i and Z_i are the fitting
parameters.

The dependences n(r) are shown graphically in Fig. 4 for Li, Na, and K atoms. The
values of r_0 for singly charged ions can be found from these dependences by postulating $n(r_0) = 1$.

The values of r_0 obtained in this way are very close to the crystallochemical radii of the
Li+ and Na+ ions but this is not true of K+. Conversely, the value of r_0 does not agree with
the position of the minimum of $\rho(r)$ in the case of LiF and NaCl but it does agree with the corresponding minimum in the case of KCl. These results can be explained without making a detailed analysis of the values of the crystallochemical radii of the ions in KCl. This explanation
runs as follows.

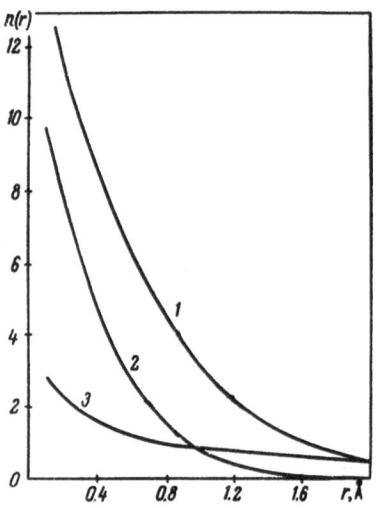

Fig. 4. Dependences n(r) for neutral atoms of K(1), Na(2), and Li (3).

The ionized state of atoms in the lattice of a solid results from a spatial and energy redistribution of the charges between these atoms. In the case of alkali halide crystals, the energy distribution is represented by the transfer of one s electron from the metal to the p state of the nonmetal. Such a transfer makes the ionic shells symmetrical and this is reflected in the diamagnetism of the ions and the solids. The transfer of one s electron should affect those parts of the f curve which are not observed in the x-ray scattering. Their different behavior in this region results in different values of the diamagnetic susceptibility and this affects the approximation of the atomic scattering factor near the zero value of $(\sin\vartheta/\lambda)$. This ionization of an atom is also accompanied by a spatial distribution of the charge. A definite short-range order and definite atomic spacings are established after ionization. The actual nature of such order and spacings depends on the shape and the size of the ions and on the interaction forces. The atomic spacings are not the sums of constants known as the "ionic radii" but depend on the nature of the two neighbors and on the interaction between them. Two cases may be encountered:

1) a large shift of an electron from the metal to the nonmetal resulting in the concentration of the excess negative charge on the surface of the anion;
2) a small shift of the charge.

In the second case, the electron clouds overlap and the excess charge is weakly localized in the overlap region. This situation is equivalent to a partial transfer of an electron from the metal to the nonmetal.

The first case is encountered in KCl. Here, the minimum of ρ (r) coincides with r_0, i.e., one electron is transferred completely to the anion. The localization of this electron near the Cl⁻ ion results in the appearance of positive energy regions in the lattice.

The second case is realized in NaCl and similar crystals, in which the minimum of ρ (r) does not correspond to r_0. The potential energy distributions in such crystals have no positive regions because there is no definite localization of the excess charge. The bonds in such crystals should be called ionic-covalent.

Literature Cited

1.   R. J. Havighurst, Proc. Nat. Acad. Sci. USA, 11:502, 507 (1925).
2.   R. J. Havighurst, Proc. Nat. Acad. Sci. USA, 12:380 (1926).
3.   R. J. Havighurst, Phys. Rev., 28:869 (1926).

4. R. J. Havighurst, Phys. Rev., 29:1 (1927).

5. J. Krug, B. Wagner, H. Witte, and E. Wölfel, Naturwiss., 40:599 (1953).

6. J. Krug, H. Witte, and E. Wölfel, Z. Phys. Chem. (Frankfurt), 4:36 (1955).

7. H. Witte and E. Wölfel, Rev. Mod. Phys., 30:51 (1958).

8. R. W. James and E. M. Firth, Proc. Roy. Soc., London, A117:62 (1928).

9. L. V. Shevtsov, V. P. Tsvetkov, and G. I. Gorokhov, in: Electronic Properties of Metals and Alloys [in Russian], Naukova Dumka, Kiev (1966).

10. G. I. Gorokhov and V. P. Tsvetkov, in: Chemical Bonds in Semiconductors and Thermodynamics, Consultants Bureau, New York (1968), p. 60.

11. G. I. Gorokhov and V. P. Tsvetkov, in: Electronic Properties of Metals and Alloys [in Russian], Naukova Dumka, Kiev (1966).

12. B. N. Vasichev, V. A. Il'ina, V. K. Latyshev, and Yu. S. Pliskin, Prib. Tekh. Eksp., No. 2, p. 51 (1960).

13. F. I. Fedorov, Kristallografiya, 10:167 (1965).

14. Ya. G. Dorfman, Diamagnetism and the Chemical Bond, American Elsevier, New York (1965).

15. J. G. Kirkwood, Phys. Z., 33:57 (1932).

16. J. C. Slater, Phys. Rev., 32:349 (1928); 36:57 (1930).

17. V. E. Lashkarev (W. E. Laschkarew) and A. S. Chaban (A. S. Tschaban), Phys. Z. Sowjetunion, 8:240 (1935).

18. H. Witte and W. Wölfel, Z. Phys. Chem. (Frankfurt), 3:296 (1955).

19. V. P. Tsvetkov and N. F. Kravtsova, Ukr. Fiz. Zh., 8:469 (1963).

20. M. Yu. Radchenko and V. P. Tsvetkov, Ukr. Fiz. Zh., 8:1364 (1963).

21. N. N. Sirota and A. U. Sheleg, Dokl. Akad. Nauk SSSR, 147:1344 (1962).

22. N. N. Sirota and A. U. Sheleg, Dokl. Akad. Nauk SSSR, 152:81 (1963).

23. N. N. Sirota and N. M. Olekhnovich, Dokl. Akad. Nauk SSSR, 148:71 (1963).

24. N. N. Sirota, Dokl. Akad. Nauk SSSR, 142:1278 (1962).

CHEMICAL AND KNIGHT SHIFTS IN $A^{III}B^{V}$ COMPOUNDS*

A. Lösche and S. Grande

Leipzig, Germany

Information was obtained on the chemical bonding in the $A^{III}B^{V}$ compounds InSb and InAs by measuring the NMR line width. It was assumed that the quadrupole effect and the chemical shift had little influence on the resonance line shift, and that the influence of the Knight shift was greater. In this way, the ionic component of the chemical bonding was determined. A theoretical analysis confirmed the experimental results.

The experimental results obtained by the nuclear magnetic resonance (NMR) method can be used to determine the nature of the chemical bonding and, particularly, the susceptibility and the carrier density. Nuclear magnetic resonance can be used to deduce information on the structure of a substance from the resonance line width and profile, from the spin–lattice relaxation, and from the shift of the resonance frequency. We shall consider some results obtained from the shift of the NMR line.

The NMR of In^{115} in InAs and InSb powders was determined by means of an autodyne detector with a stabilized frequency. The line shift was measured relative to the resonance line of In in an aqueous solution of $In(ClO_4)_3$. When both substances were in the same holder, the shift could be found directly from the calibrated separation between the lines. At low temperatures, the magnetic field was determined using one of the samples as the standard. The stability of the apparatus ensured that the error did not exceed $2 \cdot 10^{-5}$ at frequencies starting from 200 Hz.

The resonance frequency shift could be due to the quadrupole effect, the chemical shift, or the Knight shift. We shall now consider these three effects separately.

1. Q u a d r u p o l e   E f f e c t.   The spin of In^{115} is $I = 5/2$. The cubic symmetry of the lattice implies that the electric field gradient at the In nuclei vanishes if the lattice is undistorted. However, lattice defects in impurities disturb the symmetry and give rise to line broadening. The center of the line is shifted because of the second-order quadrupole effect:

$$\langle \delta v \rangle _{Av} = - \frac{v_Q^2}{16 v_L} \left[I(I+1) - \frac{3}{4} \right] \langle (1 - 3 \cos^2 \Theta)(9 \cos^2 \Theta - 1) \rangle _{Av},$$

where v_Q is proportional to the quadrupole coupling constant $eQ\varphi_{zz}$ and Θ is the angle between the external field and the direction of the coupling with the distorted nucleus. In the case of a

---

* "Crystals," pp. 208–212 (see page 3).

powder, the measured shift is the average of the angle Θ for all the nuclei oriented in different ways. It is observed at low frequencies. In the case of single crystals, we can determine the electric-field gradient at the nuclei from the angular dependence of the splitting and the shift of the NMR line.

2. Chemical Shift. The chemical shift is exhibited by all the samples. In some solids it may not be observed because of the considerable width of the NMR line.

The basic theory of the chemical shift was developed by Ramsey [1], who distinguished two components (diamagnetic and paramagnetic) of the shift

$$\sigma = \sigma^d + \sigma^p .$$

The diamagnetic component is

$$\sigma^d = \frac{e^2}{2mc^2} \langle \psi_0 \Big| \sum_i \frac{y_i^2 + z_i^2}{r_i^3} \Big| \psi_0 \rangle$$

and the paramagnetic component is

$$\sigma^p = -\frac{2e}{m^2c^2} \frac{1}{\Delta E} \langle \psi_0 \Big| \sum_{k,i} \frac{l_k l_i}{r_k^3} \Big| \psi_0 \rangle ,$$

where ΔE is the average excitation energy, whose value is difficult to determine experimentally; l_k is the orbital moment of the k-th electron; ψ_0 is the wave function of the ground state.

In the case of powders, only the isotropic part need be considered. The second of the above formulas can be transformed by introducing paramagnetic susceptibility:

$$\sigma^p = -\frac{2}{N} \chi_p \langle r^{-3} \rangle .$$

The chemical shift in solids is determined primarily by the paramagnetic component. If the total shift can be separated into the diamagnetic and the paramagnetic components, the latter component should be directly proportional to the paramagnetic susceptibility and should be a measure of the ionicity of the chemical bonding. The temperature dependence of the shift is due to the lattice vibrations and to the temperature dependence of the forbidden band width.

3. Knight Shift. This shift is observed mainly in metals and is due to the paramagnetism of the conduction electrons. The shift, measured first by Knight in 1949,

$$K = \frac{8\pi}{3} \langle |\psi_k(0)|^2 \rangle \chi_p \Omega ,$$

depends on the electron density at a nucleus and on the susceptibility of the conduction electrons. The Knight shift, due to the hyperfine exchange interaction with the conduction electrons, is observed also for semiconductors. However, the results obtained differ from the Knight shift in metals. The difference is due to the degeneracy of the energy levels in semiconductors. In the nondegenerate case, the Fermi statistics should be replaced by the Boltzmann theory. The paramagnetic susceptibility χ_p is proportional to the number of unpaired electrons; in the nondegenerate case, it is inversely proportional to the temperature. Degenerate semiconductors behave like metals.

The Knight shift should be smaller for p- than for n-type semiconductors because the eigenfunctions of the valence band of p-type materials make only a small contribution to the contact exchange interaction.

Results and Discussion

Our experimentally determined temperature dependences of the shift of the resonance line of In^{115} in InAs and InSb (containing various impurities) are presented in Fig. 1. The deviations of the experimental points from straight lines are $\pm 0.2 \cdot 10^{-4}$.

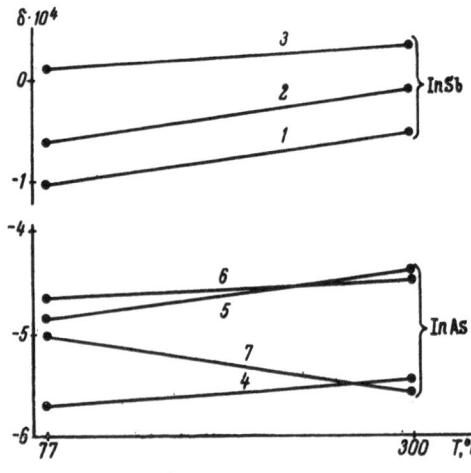

Fig. 1. Temperature dependences of the NMR line shift $\delta = (H_{In} - H_{In^{3+}})/H_{In}$ for various impurity concentrations: 1) $\sim 10^{17}$ cm^{-3}; 2) 10^{15} cm^{-3}; 3) iron impurity; 4) $\sim 4 \cdot 10^{16}$ cm^{-3}; 5) $0.7 \cdot 10^{19}$ cm^{-3}; 6) $3.4 \cdot 10^{19}$ cm^{-3}; 7) iron impurity.

We shall now try to explain the results obtained in terms of the three effects discussed in this paper.

The difference between InAs and InSb lies in the different chemical bonding. It reflects the different paramagnetic susceptibilities of these two $A^{III}B^{V}$ compounds. It follows that we can determine the ionicity of the chemical bonding in these compounds. The results obtained are in quantitative agreement with those reported by Lütgemeier [2]. A theoretical analysis of the chemical contribution to the shift was used in calculations of the absolute values and the temperature dependences of the shift for both substances. This was done by a suitable description of the bonding. The imprecise parameter representing the excitation energy was replaced, as usual, by the energy gap. The calculations were based primarily on the extensive data reported by Madelung [3].

It was found that the Knight shift was the principal component of the observed shift for the postulated chemical bonding. Several investigations had been made of the dependences of the g factor, i.e., of the susceptibility, of InAs and InSb on the temperature and on the impurity concentration [4–6]. A theoretical analysis of the influence of a considerable number of impurities and of different temperatures on the band structure, carried out by Unger [7], demonstrated that the number of unpaired electrons increases nonlinearly with increasing impurity concentration.

The conduction bands of indium arsenide and antimonide are nonparabolic. Therefore, the g factor increases strongly with increasing impurity concentration. The introduction of the necessary number of impurities causes an additional disturbance of the density of states in the conduction band. A theoretical analysis confirms the correctness of our apparently surprising experimental results. If InSb is simply a degenerate semiconductor, our results indicate a transition from the degenerate to the nondegenerate state in InAs. The temperature dependence of the Knight shift K of a nondegenerate semiconductor obeys, as mentioned already, $K \propto T^{-1}$.

The quadrupole effect should be appreciable at high impurity concentrations beginning from 10^{19} cm^{-3}. The slight asymmetry of the line shows that the second-order quadrupole effect is still small. It causes a positive shift in crystals belonging to the same group and having high impurity concentrations.

It follows from the quadrupole splitting, which increases when the temperature is raised, that the quadrupole shift should also increase with rising temperature: $\delta\nu_{Q} \propto T$. The weak dependence of the line shift on the field is evidence against a strong influence of the field on the quadrupole shift and on its temperature dependence [8].

The chemical shift depends on the temperature because the energy gap varies with the temperature. If we assume that, in the first approximation, $E = a - bT$, we find that $|\sigma^{P}| \propto T$.

These brief comments show that nuclear magnetic resonance can be used in investigations or verifications of the existing theories of chemical bonding. We expect our experimental work on A^{III}B^{V} compounds and further theoretical calculations to give quantitative data and to enable us to determine the contribution of each of the three defects to the observed shift.

The authors are grateful to D. Dietz for carrying out the experiments and to Dr. K. Unger for his valuable advice.

Literature Cited

1. N. F. Ramsey, Jr., Phys. Rev., 77:567(1950); 78:699 (1950).
2. H. Lütgemeier, Z. Naturforsch., 19a:1297 (1964).
3. O. Madelung, Physics of III-V Compounds, Wiley, New York (1964).
4. G. Römelt and D. Geist, Z. Angew. Phys., 14:99 (1962).
5. W. Zawadzki, Phys. Lett., 4:190 (1963).
6. Y. Yafet, Solid State Phys., 14:1 (1963).
7. K. Unger, Z. Naturforsch., 23a:178 (1968).
8. A. Lösche, Proc. Intern. Conf. on Magnetic Resonance and Relaxation (Coll. Ampere), Ljubljana, 1966, publ. by North-Holland, Amsterdam (1967) p. 349.

X-RAY EMISSION K SPECTRA OF SILICON
IN CHROMIUM SILICIDES*

V. P. Tsvetkov and N. D. Savchenko

K spectra of silicon were obtained in pure silicon, CrSi, and $CrSi_2$. The identity of K_{β_x} of silicon in pure silicon, CrSi, and $CrSi_2$ can be assumed from a comparison of K_{β_x} of pure silicon and chromium silicides. The semiconducting properties of chromium disilicide may be due to the presence of a partially covalent bond. A comparison of the structures of CrSi and $CrSi_2$ indicates the identity of silicon atoms in the first coordination sphere.

A study of the electrical and magnetic properties of chromium silicides showed [1, 2] that the presence of a complex electron spectrum in these alloys must be assumed in order to explain the rapid change in the electrical conductivity and magnetic susceptibility with temperature. For example, Nikitin [2] considers that the silicide spectrum consists of overlapping bands, the degree of overlap of which changes with temperature and with the lattice parameters. The nature of the overlap leads to the formation of an energy gap, the width of which is constant over a certain temperature range. This explains many of the physical properties, but the reason for the constancy of the gap width in spite of continuous increase in the lattice parameters up to the melting point is not clear. The latter fact indicates an unusual complexity of the energy spectrum of electrons in silicides and requires further investigation. In particular, it would be advisable to study the x-ray spectra associated with the outer electrons of atoms of the components of these alloys.

Studies of the x-ray absorption spectra of chromium in silicides [3] have shown that on passing from one silicide to another the fundamental absorption edge of chromium changes; this is due to a change in the nature of the silicon-atom environment of the chromium atoms. On passing from lower silicides to higher silicides, the number of Cr−Cr and Cr−Si bonds changes and there is a predominantly Si−Si bond in $CrSi_2$. Atoms of the transition element in compounds rich in silicon are further apart; this reduces the degree of overlap of the 3d−4s−4p bands.

The structure of CrSi is formed by seven-vertex configurations of Si atoms, in the centers of which are chromium atoms. Each silicon atom is surrounded by six silicon atoms at equal distances and by three sorts of chromium atoms at different distances from the silicon atom.

Chromium disilicide has a hexagonal cell made up of close-packed hexagonal layers; each Si atom is surrounded by four silicon and six chromium atoms. By analyzing the nature of the

---

* "Semiconductors," pp. 93−95 (see page 3).

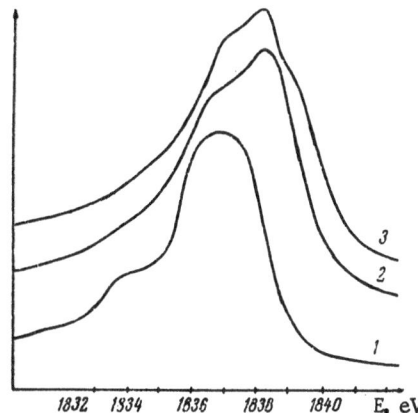

Fig. 1. $K_{\beta_{1,x}}$ bands of silicon in pure silicon (1), in $CrSi_2$ (2), and in CrSi (3).

surroundings of the silicon atoms in the two structures, it can be concluded that this changes only slightly from one structure to another. The absorption spectra of CrSi and $CrSi_2$ confirm this assumption, since the structure of the fundamental absorption edge is the same for both compounds.

Nemnonov [3] considers that the high-intensity initial absorption is due to overlapping of the chromium atom band with the 3s–3p band of the silicon atom.

A more precise idea of the electron energy spectrum in silicides was obtained from our study of the spectrum of the other component of the compound, silicon.

Electrolytic chromium and semiconducting silicon were used as the starting materials for preparing the compounds, CrSi and $CrSi_2$. The alloys were prepared by fusion in an induction furnace filled with argon. The samples obtained, of stoichiometric composition, were subjected to chemical and phase analyses. Because the alloys were very brittle, it was not possible to prepare plates. For this reason, the samples were ground to a powder and rubbed into the grooved surface of the anode.

The study was carried out by the primary and, to test the oxidation, by the secondary method on a DRS–2. The analyzer was a quartz crystal with a (10$\bar{1}$0) prism plane. The reference lines used were the $K_{\alpha_{1,2}}$ Ca doublet in the fourth order of reflection and $K_{\alpha_{1,2}}$ Ca in the second. The tube operating conditions were 4 kV and 20 mA.

Spectrograms were processed on an MF–4 point-by-point; after averaging 4 or 5 spectrograms, curves were plotted of intensity distribution along the band.

The resolution in the $K_{\beta_{1,x}}$ region of silicon is 15,000; in the $K_{\alpha_{1,2}}$ region, it is 12,000.

Curves are shown in Fig. 1 for the intensity of the $K_{\beta_{1,x}}$ bands for pure silicon, chromium silicide, and chromium disilicide. It has been previously shown [4, 5] that in the silicides of nickel, iron, and cobalt the shape of the $K_{\beta_{1,x}}$ band of Si and its energy position undergo appreciable changes compared with pure silicon, depending on the phase structure. The curves of the $K_{\beta_{1,x}}$ bands of chromium silicides, obtained in this work, differ only slightly from the pure-silicon curve. Also characteristic of chromium silicides is the fact that there is a negligible change of the initial absorption region of chromium in these compounds.

These facts indicate that the structures of CrSi and $CrSi_2$ are very similar. In these compounds, the silicon atoms in the first coordination sphere are mainly surrounded by silicon; the chromium atoms are much further away. Such surroundings make it possible to speak of mainly covalent Si–Si bonds being present, similar to the bonds in pure silicon.

The participation of the outer electrons in the chemical bonds, accompanied by a change in the spatial electron density distribution, has an effect on the position of the K level. As Karal'nik [6] showed, the level displacement caused by a change in the screening of the nucleus by the valence electrons is as high as 1 eV. Our investigation of the position of the $K_{\alpha_{1,2}}$ line of silicon in the chromium silicides showed that, within the limits of experimental error, it coincides with its position in the pure element.

The position and intensity of the satellites $K_{\alpha_{3,4}}$ were constant. It should therefore be concluded that the whole x-ray spectrum of silicon in chromium silicides differs only slightly from that in the pure element. This is obviously due to the fact that silicon atoms in chromium silicides are bound by mainly covalent bonds, similar to the bonds in the pure element.

Literature Cited

1. I. Z. Radovskii, T. S. Shubina, P. V. Gel'd, and F. A. Sidorenko, Porosh. Met., No. 2, p. 33 (1965).
2. E. N. Nikitin, Zh. Tekh. Fiz., 28:23 (1958).
3. S. A. Nemnonov and A. Z. Men'shikov, Fiz. Metal. Metalloved., 9:385 (1960).
4. R. D. Bartosevich, N. D. Savchenko, and V. P. Tsvetkov, Ukr. Fiz. Zh., 11:424 (1966).
5. V. P. Tsvetkov, N. D. Savchenko, and M. E. Radchenko, in: Electronic Structure of Metals and Alloys [in Russian], Naukova Dumka, Kiev (1965).
6. S. M. Karal'nik, Ukr. Fiz. Zh., 10:913 (1965).

ELECTRON DIFFRACTION STUDY OF SODIUM BROMIDE THIN FILMS*

A. G. Buntar' and A. F. Margolina

Data from an electron diffraction measurement of integrated intensities of Debye reflections for polycrystalline sodium bromide films are used to construct curves of the electron atomic scattering factors of Na and Br. An approximation is given of experimental f curves by analytical expressions obtained in the Slater approximation. The parameters of the approximation are used to calculate the charge on the atoms and also to determine the potential distribution in the sodium bromide unit cell.

The physical properties of atoms and the interaction between them are determined by their electron structure. Because of this, it is important to determine experimentally the spatial distribution of the electron density and the potential, the variation of these properties when chemical compounds are formed and when different materials crystallize, and also the excess charge on the atoms arising from such a redistribution of electrons.

Several authors [1, 2] have shown that a direct experimental determination of the spatial distribution of the electron density and potential is possible not only by the x-ray diffraction methods, which are widely used, but also by the electron diffraction methods. Also, when investigating certain problems, for example, in the study of ionic compounds, electron diffraction methods of analysis are more promising in view of the profound effect of atomic ionization on the curves of the atomic scattering factor.

Experimental Method and Results

We carried out an electron diffraction study of sodium bromide. This was chosen as the subject for study because of its simple crystal lattice (K–5 type), the stability of the chemical compound in evaporation and condensation, and also the fairly high value of its lattice constant ($a = 5.975$ Å), enabling us to obtain information on the shape of the form factor curves at low scattering angles.

Sodium bromide films were obtained by vacuum evaporation directly in the electron diffraction camera (ÉG–1) in a vacuum of approximately $5 \cdot 10^{-4}$ mm Hg; amorphous celluloid films were used as the substrates. When they condensed, the NaBr crystallites tended to form grains. It was noted that the preferred orientation of the crystallites depended on the film thickness. When the film thickness was reduced, the preferred orientation disappeared. The thickness of

---

* "Semiconductors," pp. 96–105 (see page 3).

the sodium bromide films was obtained by a calculation based on the geometry of the vapor-deposition equipment and the weight of the material dispersed; the thickness was approximately 100 to 150 Å.

The absence of texture in the films was determined by photometrically scanning the oblique electron diffraction patterns (exposure angles 30 and 50°) in different azimuthal directions with a 30° interval. Only electron diffraction patterns without any signs of texture were used for the measurements. A study of the oblique exposures also showed that the absorption coefficient was almost zero, which made it possible to disregard absorption of the electron beam in further investigations (the method of determining the absorption coefficient is given in detail in [3]).

The reflection intensities were determined from a series of electron diffraction patterns with multiple exposures, using the microscanning method described in [1, 4]. The results were averaged from the data for two samples; the average accuracy of intensity determinations varied from 2% for strong reflections to 10% for weak reflections.

It has been shown [1] that the criterion for changing from kinematic scattering to dynamic scattering is given by the size of the crystallites A, determined from the equation

$$\lambda \left| \frac{\Phi_{hkl}}{\Omega} \right| A \leqslant 1. \tag{1}$$

Vainshtein [1] estimates theoretically the critical thickness of the crystallites, which was about 200 Å for Al and about 50 Å for Au. These results were later confirmed experimentally by Yazmin and Pinsker [5].

The sizes of NaBr crystallites, determined from the line widths, were approximately 40 to 60 Å, which made it possible, in changing from the detected intensities I_{hkl} to structure amplitudes Φ_{hkl}, to use the kinematic scattering approximation, according to which, for poly-crystalline samples, we have

$$|\Phi_{hkl}| \sim \sqrt{I_{hkl}/d_{hkl}^2 p}, \tag{2}$$

where d_{hkl} is the interplanar spacing and p is the multiplicity factor.

The values obtained for the structure amplitudes decrease more rapidly with increase in $\sin\vartheta/\lambda$ than the theoretical values, i.e., the temperature factor is having an effect; this also indicates the kinematic nature of scattering on these samples.

Values of the temperature factor and the normalization factor K, making it possible to change to p units, were obtained from a logarithmic plot $\ln(\Phi_{theor}/\Phi_{exp}) = F(\sin^2\vartheta/\lambda^2)$ [6, 7]. Form factors for neutral sodium and bromine atoms were used to calculate Φ_{theor}; these values are tabulated in [1] and [8], respectively. The value of B determined in this manner was 0.525 Å^2, and K = 4.58.

Analysis of the data in Table 1 shows that, for $\sin\vartheta/\lambda \geq 0.37$, the experimental values of the structure amplitudes agree, within the limits of experimental error, with theoretical values. At small scattering angles, the experimental values of Φ_{hkl} are lower than the theoretical values; this may be due to the ionic nature of the bonds in the NaBr lattice. The appreciable difference between Φ_{exp} and Φ_{theor} for the (200) and (400) reflections should be noted; the reason for this may be the anisotropy of the thermal vibrations in the NaBr lattice.

By interpolating the experimental curves of the sum and difference of atomic factors, we separated them and plotted curves of atomic scattering factors of sodium and bromine (Fig. 1). Atomic factors for the Na$^+$ ion were taken from [1], and for the Br$^-$ ion from [9]. It can be seen that the theoretical curves for the Na$^+$ ion and the neutral Na atom differ only slightly from one another over the investigated range of $\sin\vartheta/\lambda$; the experimental values of f_{Na} are

TABLE 1. Values, Corrected for the
Experimental Temperature (in p units),
of the Determined Structure Amplitudes
and Theoretical Φ_{hkl} Obtained from
Electron-Scattering Form Factors of
Neutral Na and Br Atoms

hkl	$\dfrac{\sin\vartheta}{\lambda}$	$\left(\dfrac{\Phi_{hkl}}{4}\right)_{theor}$	$\left(\dfrac{\Phi_{hkl}}{4}\right)_{exp}$
111	0.145	1.36	1 31 ± 0.02
200	0.167	2.94	2.36 ± 0.01
220	0.237	2.28	2.15 ± 0.01
311	0.278	0.89	0.89 ± 0.01
222	0.290	1.82	1.82 ± 0.03
400	0.335	1.60	1.26 ± 0.03
331	0.368	0.66	0.68 ± 0.02
420	0.374	1.40	1.38 ± 0.02
422	0.410	1.27	1.25 ± 0.02
511,333	0.435	0.52	0.56 ± 0.01
531	0.490	0.44	0.45 ± 0.02
600,442	0.500	0.99	0.98 ± 0.02
620	0.529	0.92	0.92 ± 0.02
533	0.549	0.38	0.39 ± 0.05
622	0.556	0.86	0.86 ± 0.03
711	0.597	0.34	0.40 ± 0.06
640	0.604	0.76	0.76 ± 0.06
642	0.625	0.73	0.72 ± 0 03
820,644	0.690	0.62	0.63 ± 0.06

somewhat different from the theoretical values and at low sin ϑ/λ do not tend to climb as steeply as the theoretical values. For bromine, there is an appreciable difference between the theoretical curves of the atomic factors for the neutral bromine atom and the Br⁻ ion; the experimental values lay between the two.

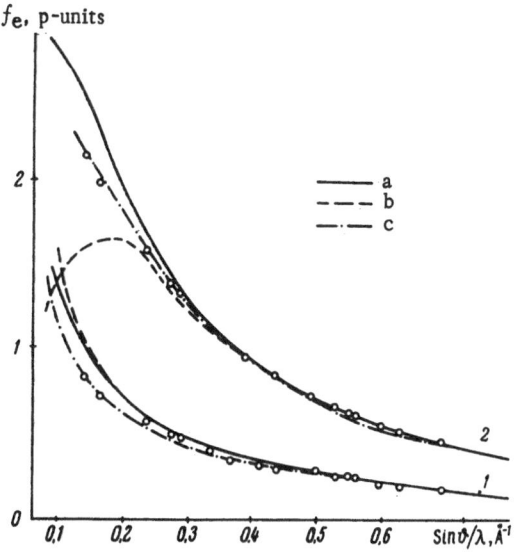

Fig. 1. Experimental and theoretical values of atomic factors of electron scattering for sodium (1) and bromine (2): a) for neutral atoms; b) for ions; c) fitted curves.

Determination of the Spatial Distribution of Charge and Potential

The determination of the charge on the atoms and the electron density and potential distribution in the NaBr lattice was carried out by approximation of the experimental f curves, using analytical expressions for the form factor of electron scattering in the Slater approximation, which are given in [10].

According to the proposed approximation, the expressions for the atomic factors representing electron scattering by neutral atoms can be obtained by summation of the components of the atomic factor $f_{nlm}(\mathbf{s})$ over the quantum states occupied by electrons, $f_{nlm}(\mathbf{s})$ being a function of the screening coefficient γ and the effective quantum number n. Expressions for the components of the atomic factors for K, L, and M shells are given in [10]. For the N shell, for which

Slater proposes an effective quantum number n = 3.7, the atomic factor can only be presented in the form of a sum with an infinite number of components. The series may be terminated if the effective quantum number for the N shell is taken as 3.5, 4.0, or 4.5. We calculated values of the atomic factor for the neutral Br atom with different values of n. The most satisfactory agreement with the theoretical form factors, calculated according to the Thomas−Fermi−Dirac model, was obtained at n = 4.5; screening coefficients proposed in [11] were used in the calculations. The equation of the atomic scattering function for the N shell in the case of a spherically symmetrical electron density distribution and n = 4.5 has the following form:

$$f_p(s) = \frac{N_{nlm}}{(1 + \varkappa_{nl}^2 s^2)^9} \left[1 - \frac{28}{3} \varkappa_{nl}^2 s^2 + 14 \varkappa_{nl}^4 s^4 - 4 \varkappa_{nl}^6 s^6 + \frac{1}{9} \varkappa_{nl}^8 s^8 \right],$$

(3)

$$f_e(s) = \frac{k N_{nlm} \varkappa_{nl}^2}{(1 + \varkappa_{nl}^2 s^2)^2} \left[\frac{256}{9} - \frac{64}{3} a_{nl} + \frac{16}{3} a_{nl}^2 + \frac{8}{9} a_{nl}^3 + a_{nl}^4 + a_{nl}^5 + a_{nl}^6 + a_{nl}^7 + a_{nl}^8 \right],$$

where

$$s = 4\pi \sin \vartheta / \lambda,$$

$$a_{nl} = 1 + \varkappa_{nl}^2 s^2,$$

$$\varkappa_{nl} = \frac{1}{\eta_{nl}} = \frac{n a_0}{2(Z - \gamma)}.$$

(4)

Here, a_0 is the radius of the first Bohr orbit; n is the effective quantum number; Z is the atomic number of the element; γ is the screening coefficient.

In the case of ions the approximate equations have the form

$$f_e^+(s) = \sum_{nlm} f_{nlm}(s) + \frac{q}{s^2},$$

$$f_e^-(s) = \sum_{nlm} f_{nlm}(s) - \frac{f_p'(s)}{s^2},$$

(5)

where q is the charge on the ions; $f_p'(s)$ is the atomic factor in the case of x-ray scattering given in Eq. (3) (it is determined by the number of electrons q attached to a neutral atom).

To carry out the described approximation, the number of electrons in each quantum state must be determined and also the numerical values of the approximation parameters $\varkappa_{nl}$. Assuming that the state of the inner-shell electrons is practically unchanged during crystal formation, and using the screening coefficients for neutral atoms given in [11], we determined $\varkappa_{nl}$ for the inner-shell electrons of the Na^+ and Br^- ions. The approximation parameters and the number of electrons in the outer layers were determined by mathematical analysis of the experimental data. By using the method of averages [12], we were able to determine the coefficients $\varkappa_{4p}$ for bromine, $\varkappa_{3s}$ for sodium, and also the number of electrons in these states. To do this we solved a system of equations of the type

$$\sum_i \varepsilon(s_i) = 0,$$

(6)

where $\varepsilon(s_i)$ is the difference between the experimental and theoretical values of the atomic factor.

Using this method, the following parameters of the approximation $\varkappa_{nl}$ were determined:

Atoms	Quantum state							
	$1s$	$2s$	$2p$	$3s$	$3p$	$3d$	$4s$	$4p$
Na	0.0248	0.0789	0 0779	0.1862				
Br	0.0077	0,0206	0.0170	0.0392	0.0405	0.0405	0.1002	0.1040

Our calculation showed that in NaBr the bromine atom gains and the sodium atom loses about 0.8 electron. The results obtained were used to plot form factor curves for bound sodium and bromine atoms in the compound NaBr. The results are shown in Fig. 1 by the chain lines; it can be seen that there is fairly good agreement between the approximated curves and the experimental form factor values.

The radial electron density distribution $u(r) = 4\pi r^2 \rho(r)$ can be estimated on the basis of the data obtained.

In the Slater approximation,

$$\rho(r) = R_{nl}(r) = \frac{N_{nlm}\eta_{nl}^{2n+1}}{2n!}\, r^{2n-2}\exp(-\eta_{nl}r). \tag{7}$$

Curves of $u(r)$, calculated using the parameters given above for bound sodium and bromine ions, are shown in Fig. 2, where, for comparison, the electron density distribution for neutral Na and Br atoms is also given; the latter was calculated using the parameters η_{nl} taken from [11]. It should be noted that in the bound Br atom some compression of the 4p shell is due to the excess charge from the Na atom. However, the electron density does not fall to zero between the sodium and bromine atoms in the NaBr lattice, and a small (0.5 el/Å) electron bridge is clearly noticeable. The ionic radii of Na$^+$ and Br$^-$ were estimated from the minimum value of the electron density; they were 1.06 and 1.93 Å, respectively.

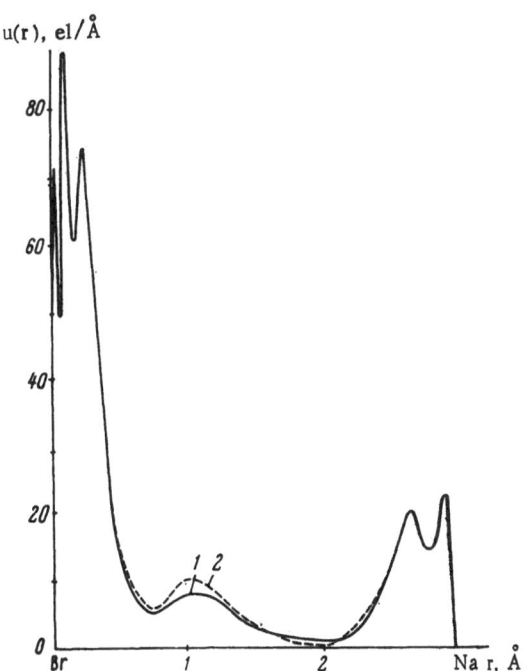

Fig. 2. Radial electron density distribution
$u(r) = u_{Br}(r) + u_{Na}(r)$ for neutral atoms
(1) and bound ions (2) in the NaBr lattice.

The data which we obtained in the approximation can be used to find the potential distribution in the NaBr lattice. The equations from which the potential of the atom can be calculated in the Slater approximation are given in [13]. In our case, if we accept that the charge on the atoms in the NaBr crystal is 0.8, the equations for the ion potentials have the form

$$\Phi\,(\mathbf{r})_{Na^+} = \frac{0.8e}{r} + e\sum_{nl} N_{nl}\exp\left(-\eta_{nl}r\right)\left[\frac{1}{r} + \sum_{p=1}^{2n-1}\eta_{nl}^{2n-p}\,r^{2n-p-1}\,\frac{p}{2n\,(2n-p-1)!}\right],$$

$$\Phi\,(\mathbf{r})_{Br^-} = -\frac{0.8e}{r} + e\sum_{nl} N_{nl}\exp\left(-\eta_{nl}\,r\right)\left[\frac{1}{r} + \sum_{p=1}^{2n-1}\eta_{nl}^{2n-p}\,r^{2n-p-1}\,\frac{p}{2n\,(2n-p-1)!}\right]. \tag{8}$$

An analysis of these equations showed that the second terms in the expressions (8) rapidly decrease as r increases. When calculating the lattice potential due to these terms, the treatment could therefore be limited at various points to only the nearest neighbors (first coordination sphere).

The method proposed in [14] was used to calculate the potential distribution in the lattice. The contribution due to point ions in the overall lattice potential was calculated on a Minsk–14 computer. We determined the potential distribution in the crystal lattice along the main crystallographic directions. The results of these calculations are shown in Fig. 3, from which it can be seen that the potential distribution in the sodium bromide lattice is anisotropic. The potential distribution in the [100] direction is unsymmetrical. The ranges of the attraction forces for ions in the lattice were determined from the minimum value of the potential curve; for sodium and bromine ions, these radii were 1.19 and 1.79 Å, respectively. Ionic radii (in Å) shown in the table below for sodium and bromine were determined experimentally from the data on the elec-

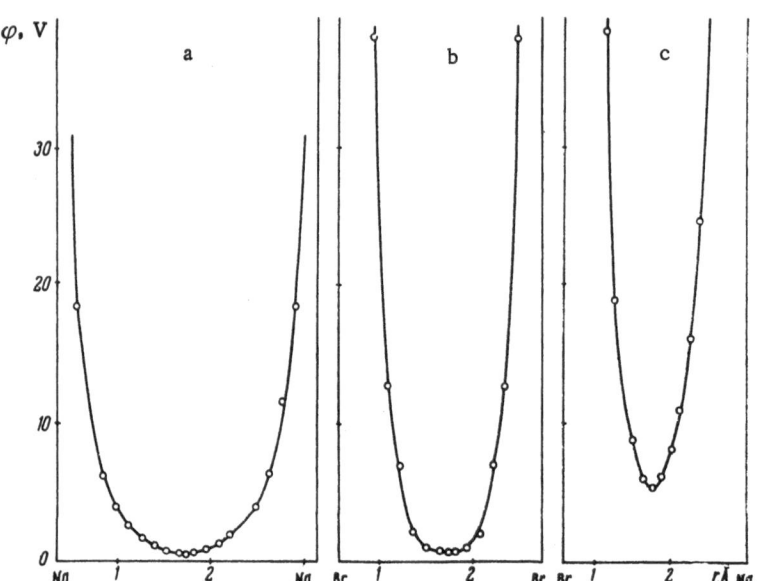

Fig. 3. Potential distribution in the directions Na–Na (a),
Br–Br (b), Na–Br (c).

tron density and potential distributions and taken from various sources in the literature:

Ions	Radii, Å				
	from Fig. 2	from Fig. 3c	from [15]	from [16]	from [17]
Na	1.06	1.19	0.95	0.98	1.58
Br	1.93	1.79	1.95	1.96	1.14

It can be seen that the ionic radii determined experimentally from distribution plots of u(r) and φ(r) are very similar and are in good agreement with theoretical data of Pauling [15] and Goldschmidt [16], whereas according to data in [17] the ionic radius of Na^+ is greater than that for Br^-.

Literature Cited

1. B. K. Vainshtein, Structure Analysis by Electron Diffraction, Pergamon, Oxford (1964).
2. N. N. Sirota, in: Chemical Bonds in Semiconductors and Thermodynamics (ed. by N. N. Sirota), Consultants Bureau, New York (1968), p. 1.
3. A. G. Buntar', V. V. Trofimova, and L. A. Shchepilova, in: Chemical Bonds in Semiconductors and Solids (ed. by N. N. Sirota), Consultants Bureau, New York (1967), p. 285.
4. É. E. Vainshtein, Methods of Quantitative Analysis by X-Ray Spectroscopy [in Russian], Izd. Akad. Nauk SSSR, Moscow (1956).
5. I. I. Yazmin and Z. G. Pinsker, Dokl. Akad. Nauk, SSSR, 65:645 (1949).
6. S. Togawa, J. Phys. Soc. Jap., 19:1696 (1964).
7. M. P. Arbuzov, E. T. Kochkovskaya, and B. V. Khaenko, in: Chemical Bonds in Semiconductors and Thermodynamics (ed. by N. N. Sirota), Consultants Bureau, New York (1968), p. 45.
8. L. I. Mirkin, Handbook of X-Ray Analysis of Polycrystalline Materials, IFI/Plenum Press, New York (1964).
9. A. G. Buntar' and V. V. Storozhenko, in: Energy Spectra of Electrons in Metals [in Russian], Naukova Dumka, Kiev (1965), p. 112.
10. A. G. Buntar', Kristallografiya, 11: 722 (1966).
11. E. Clementi and D. L. Raimondi, J. Chem. Phys., 38:2686 (1963).
12. K. P. Yakovlev, Mathematical Analysis of Experimental Results [in Russian], GITTL, Moscow (1953).
13. A. G. Buntar', in: Chemical Bonds in Semiconductors and Thermodynamics (ed. by N. N. Sirota), Consultants Bureau, New York (1968), p. 51.
14. K. I. Gorokhov and V. P. Tsvetkov, in: Chemical Bonds in Semiconductors and Thermodynamics (ed. by N. N. Sirota), Consultants Bureau, New York (1968), p. 60.
15. L. Pauling, The Nature of the Chemical Bond, 3rd. ed., Cornell University Press, Ithaca, N. Y. (1960).
16. V. Goldschmidt, Fortschr. Min., 15:75 (1931).
17. S. S. Batsanov and V. I. Durakov, Zh. Strukt. Khim., 1:353 (1960).

ANISOTROPY OF THE SCATTERING OF X RAYS AND ELECTRONS BY CUBIC CRYSTALS*

A. G. Buntar'

An analysis is given of the influence of perturbing fields of the O_h and T_d symmetry on the x-ray and the electron scattering form factors. The results obtained are used to calculate the structure amplitudes for bcc and fcc lattices. It is shown that the anisotropy of the x-ray or the electron scattering in transition metals should be observed in single-crystal and polycrystalline samples. The changes in the relative intensities in the x-ray diffraction patterns resulting from deviations of the spatial electron distribution from the central symmetry may reach 20–40% for elements in the second period and 4–5% for transition metals. The corresponding changes in the electron diffraction patterns should be somewhat smaller. Thus, for transition metals, the changes in the electron diffractograms should be about 1%.

When x-ray or electron diffraction methods are used to study the nature of the binding forces in a crystal, it is desirable to know the influence on the scattering power of atoms of those changes in the spatial distribution of electrons which are due to the interaction between the particles forming the crystal lattice.

The present paper is concerned with the problem of the influence of perturbing fields of the O_h or T_d symmetry on the scattering form factor (atomic scattering factor) of x rays and electrons. The results obtained are used to calculate the structure amplitudes of bcc and fcc lattices. Possible distortions of the wave functions resulting from the influence of the crystal field are ignored.

Scattering Form Factors of X Rays and Electrons

In the one–electron approximation, the x–ray form factor may be represented by

$$f(s) = \sum_k N_k \cdot f_k(s), \tag{1}$$

where N_k is the number of electrons in the k-th quantum state; $f_k(s)$ is the component of the form factor corresponding to the state,

$$f_k(s) = \int \psi_k(r) \cdot \exp[i(sr)] \cdot \psi_k^*(r) \cdot d\tau, \tag{2}$$

where $\psi_k(r)$ is the relevant wave function.

In a centrally symmetric field, there are $(2l + 1)$ wave functions for every value of the

---

*"Semiconductors," pp. 106–115 (see page 3).

quantum number l. These $(2l + 1)$ wave functions form the basis of a reducible representation for the cubic (group O_h) and the tetrahedral (group T_d) fields since the maximum degrees (dimension) of the irreducible representations of these groups is 3.

The wave functions ψ_k, forming the basis of the irreducible representations, can be represented in the form of a linear combination of wave functions ψ_j obtained by solving Schrödinger's equation for a centrally symmetric field:

$$\psi_k(\mathbf{r}) = \sum_j C_{kj} \cdot \psi_j(\mathbf{r}). \tag{3}$$

Substituting Eq. (3) into Eq. (2), we obtain

$$f_k(\mathbf{s}) = \sum_{i,j} C_{ik} \cdot C_{jk} \cdot f_{ij}(\mathbf{s}), \tag{4}$$

where

$$f_{ij}(\mathbf{s}) = \int \psi_i(\mathbf{r}) \cdot \exp\left[i(\mathbf{sr})\right] \cdot \psi_j^*(\mathbf{r}) \cdot d\tau. \tag{5}$$

A wave function ψ_j can be represented in the form

$$\psi_j(\mathbf{r}) = \sqrt{\frac{2l_j + 1}{4\pi} \cdot \frac{(l_j - m_j)!}{(l_j + m_j)!}} \; P_{l_j}^{m_j}(\cos\Theta) R_{n_j l_j}(r) \exp\left[i m_j \varphi\right]. \tag{6}$$

The radial part of this wave function, R_{nl}, can be obtained using Slater's approximation [1]

$$R_{nl}(r) = \left[\frac{\eta_{nl}^{2n+1}}{\Gamma(2n+1)}\right]^{1/2} r^{n-1} \cdot \exp\left[-\frac{\eta_{nl} r}{2}\right], \tag{7}$$

where $\eta_{nl} = \dfrac{1}{\varkappa_{nl}} = \dfrac{2(Z-\gamma)}{n a_0}$; γ is the screening coefficient; Z is the atomic number; a_0 is the radius of the first Bohr orbit.

Substituting Eqs. (6) and (7) into Eq. (5) and integrating, we obtain

$$f_{ij}(\mathbf{s}) = A_{ij} \cdot \exp\left[i\mu\gamma\right] \cdot \sum_{t=|l_i-l_j|}^{l_i+l_j} B_t P_t^\mu(\cos\beta) \times$$

$$\times \frac{s^t \eta_i^{n_i+1/2} \eta_j^{n_j+1/2}}{(\eta_i + \eta_j)^{n_i+n_j+t+1}} F\left(\frac{n_i + n_j + t + 1}{2}, \frac{n_i + n_j + t + 2}{2}; t + \frac{3}{2}; -\frac{4s^2}{(\eta_i + \eta_j)^2}\right), \tag{8}$$

where

$$A_{ij} = (-1)^{m_i} \sqrt{\frac{(2l_i + 1)(2l_j + 1)}{\Gamma(2n_i + 1)\Gamma(2n_j + 1)}} \; ; \tag{9}$$

$$B_t = i^t (2t + 1) \sqrt{\frac{(t - \mu)!}{(t + \mu)!}} \; \frac{2^{n_i+n_j+t+1} \Gamma(n_i + n_j + t + 1)}{(2t + 1)!!} \begin{pmatrix} l_i & l_j & t \\ -m_i & m_j & \mu \end{pmatrix} \begin{pmatrix} l_i & l_j & t \\ 0 & 0 & 0 \end{pmatrix}; \tag{10}$$

$P_t^\mu(x)$ are the associated Legendre functions; $F(a, b; c; z)$ is the hypergeometric function; s, β, and γ are the spherical coordinates of the vector $\mathbf{s}$; $\mu = |m_i - m_j|$; $\begin{pmatrix} l_1 & l_2 & l_3 \\ m_1 & m_2 & m_3 \end{pmatrix}$ are the 3j Wigner symbols.

A. G. BUNTAR'

TABLE 1. Scattering Form Factor of X Rays in Fields of O_h and T_d Symmetry

n	l	Γ	ψ_k	f_k (s)
2	1	δ	x	$f(2p) + V(2p)[2P_2(\cos\beta) - \cos2\gamma \cdot P_2^2(\cos\beta)]$
			y	$f(2p) + V(2p)[2P_2(\cos\beta) + \cos2\gamma \cdot P_2^2(\cos\beta)]$
			z	$f(2p) - 4V(2p)P_2(\cos\beta)$
3	1	δ	x	$f(3p) + \dfrac{4}{15}D(3p)[2P_2(\cos\beta) - \cos2\gamma \cdot P_2^2(\cos\beta)]$
			y	$f(3p) + \dfrac{4}{15}D(3p)[2P_2(\cos\beta) + \cos2\gamma \cdot P_2^2(\cos\beta)]$
			z	$f(3p) - \dfrac{16}{15}D(3p)P_2(\cos\beta)$
	2	γ	$z^2 - r^2$	$f(3d) - \dfrac{16}{21}D(3d)P_2(\cos\beta) + \dfrac{96}{7}V(3d)P_4(\cos\beta)$
			$x^2 - y^2$	$f(3d) + \dfrac{16}{21}D(3d)P_2(\cos\beta) + \dfrac{2}{7}V(3d)[8P_4(\cos\beta) + \dfrac{1}{3}\cos4\gamma \cdot P_4^4(\cos\beta)]$
			xy	$f(3d) + \dfrac{16}{21}D(3d)P_2(\cos\beta) + \dfrac{2}{7}V(3d)[8P_4(\cos\beta) - \dfrac{1}{3}\cos4\gamma \cdot P_4^4(\cos\beta)]$
		ε	yz	$f(3d) - \dfrac{4}{21}D(3d)[2P_2(\cos\beta) - \cos2\gamma \cdot P_2^2(\cos\beta)] - \dfrac{16}{7}V(3d)[4P_4(\cos\beta) + \dfrac{1}{3}\cos2\gamma \cdot P_4^2(\cos\beta)]$
			xz	$f(3d) - \dfrac{4}{21}D(3d)[2P_2(\cos\beta) + \cos2\gamma \cdot P_2^2(\cos\beta)] - \dfrac{16}{7}V(3d)[4P_4(\cos\beta) - \dfrac{1}{3}\cos2\gamma \cdot P_4^2(\cos\beta)]$

It follows from Eq. (10) that $B_t \neq 0$ if $|l_i - l_j| \leq t \leq l_i + l_j$ and $l_i + l_j + t = 2g$, where g is an integer.

We shall ignore the possible hybridization of the states with different values of n and l. In this case, we may assume that $n_i = n_j = n$ and $l_i = l_j = l$.

It was demonstrated by von der Lage and Bethe [2] that the irreducible representation for the p states ($l = 1$) in the cubic field is the three-dimensional representation δ, which corresponds to the cubic harmonics of the x, y, and z type. If $l = 2$, the irreducible representations are a two-dimensional representation γ with the basis functions $z^2 - r^2$ and $x^2 - y^2$, and a three-dimensional representation ε whose basis functions are of the type xy, yz, and xz.

The same irreducible representations apply to the group T_d. Thus, we can easily determine the coefficients C_{kj} and this allows us to find expressions for the scattering form factor of x rays by means of Eqs. (1), (4), and (8)–(11).

The formulas for the components of the form factor corresponding to the 2p, 3p, and 3d states are listed in Table 1. The following notation is used in this table:

$$f(2p) = \frac{1 - \varkappa_{21}^2 s^2}{(1 + \varkappa_{21}^2 s^2)^4} \; ; \quad V(2p) = \frac{\varkappa_{21}^2 s^2}{(1 + \varkappa_{21}^2 s^2)^4} \; ; \quad f(3l) = \frac{1 - \dfrac{10}{3}\varkappa_{3l}^2 s^2 + \varkappa_{3l}^4 s^4}{(1 + \varkappa_{3l}^2 s^2)^6} \; ; \quad D(3l) = \frac{\varkappa_{3l}^2 s^2 (7 - 3\varkappa_{3l}^2 s^2)}{(1 + \varkappa_{3l}^2 s^2)^6} \; ;$$

$$V(3d) = \frac{\varkappa_{32}^4 s^4}{(1 + \varkappa_{32}^2 s^2)^6} \; .$$

The expression for the scattering form factor of electrons $\varphi(\mathbf{s})$ can be obtained if we use the expressions just derived and the relationship between the scattering form factors of x rays and electrons

$$\varphi(\mathbf{s}) = k\,\frac{Z - f(\mathbf{s})}{s^2}. \tag{11}$$

If we measure $\varphi(\mathbf{s})$ in p units [3], we find that $k \approx 1.58$. The components of the scattering form factor of the electrons corresponding to the 2p, 3p, and 3d states are given in Table 2, where the following notation is employed:

$$a_{nl} = 1 + \varkappa_{nl}^2 s^2.$$

Anisotropy of Structure Amplitudes

The expressions obtained for the form factors can be used to determine the structure amplitudes of cubic lattices, making allowance for the anisotropy in the scattering of x rays and electrons.

It is known that when x rays or electrons are scattered by a crystal a reflection is obtained if the following condition is satisfied:

$$\mathbf{s} = 2\pi\,\mathbf{H}, \tag{12}$$

where $\mathbf{H}$ is the reciprocal lattice vector.

This makes it possible to deduce the following relationship between the spherical coordinates of the vector $\mathbf{s}$ and the reflection indices (h, k, l), which applies to cubic crystals:

$$\cos\beta = \frac{2\pi l}{as} = \frac{l}{\sqrt{h^2 + k^2 + l^2}}, \quad \cos\gamma = \frac{h}{\sqrt{h^2 + k^2}}, \tag{13}$$

where a is the lattice period.

TABLE 2. Scattering Form Factor of Electrons in Fields of O_h and T_d Symmetry

n	l	Γ	ψ_k	$\varphi_k(s)$
2	1	δ	x	$\varkappa_{21}^2\left\{2 + a_{21} + a_{21}^2 + a_{21}^3 - 2P_2(\cos\beta) + \cos 2\gamma \cdot P_2^2(\cos\beta)\right\}/a_{21}^4$
			y	$\varkappa_{21}^2\left\{2 + a_{21} + a_{21}^2 + a_{21}^3 - 2P_2(\cos\beta) - \cos 2\gamma \cdot P_2^2(\cos\beta)\right\}/a_{21}^4$
			z	$\varkappa_{21}^2\left\{2 + a_{21} + a_{21}^2 + a_{21}^3 + 4P_2(\cos\beta)\right\}/a_{21}^4$
3	1	δ	x	$\varkappa_{31}^2\left\{\dfrac{16}{3} + a_{31}^2 + a_{31}^3 + a_{31}^4 + a_{31}^5 - \dfrac{4}{15}(10 - 3a_{31})\,[2P_2(\cos\beta) - \cos 2\gamma \cdot P_2^2(\cos\beta)]\right\}/a_{31}^6$
			y	$\varkappa_{31}^2\left\{\dfrac{16}{3} + a_{31}^2 + a_{31}^3 + a_{31}^4 + a_{31}^5 - \dfrac{4}{15}(10 - 3a_{31})\,[2P_2(\cos\beta) + \cos 2\gamma \cdot P_2^2(\cos\beta)]\right\}/a_{31}^6$
			z	$\varkappa_{31}^2\left\{\dfrac{16}{3} + a_{31}^2 + a_{31}^3 + a_{31}^4 + a_{31}^5 + \dfrac{16}{15}(10 - 3a_{31})\,P_2(\cos\beta)\right\}/a_{31}^6$
	2	γ	$z^2 - r^2$	$\varkappa_{32}^2\left\{\dfrac{16}{3} + a_{32}^2 + a_{32}^3 + a_{32}^4 + a_{32}^5 + \dfrac{16}{21}(10 - 3a_{32})\,P_2(\cos\beta) - \dfrac{96}{7}(a_{32} - 1)\,P_4(\cos\beta)\right\}/a_{32}^6$
			$x^2 - y^2$	$\varkappa_{32}^2\left\{\dfrac{16}{3} + a_{32}^2 + a_{32}^3 + a_{32}^4 + a_{32}^5 - \dfrac{16}{21}(10 - 3a_{32})\,P_2(\cos\beta) - \dfrac{2}{7}(a_{32} - 1)\left[8P_4(\cos\beta) + \dfrac{1}{3}\cos 4\gamma \cdot P_4^4(\cos\beta)\right]\right\}/a_{32}^6$
			xy	$\varkappa_{32}^2\left\{\dfrac{16}{3} + a_{32}^2 + a_{32}^3 + a_{32}^4 + a_{32}^5 - \dfrac{16}{21}(10 - 3a_{32})\,P_2(\cos\beta) - \dfrac{2}{7}(a_{32} - 1)\left[8P_4(\cos\beta) - \dfrac{1}{3}\cos 4\gamma P_4^4(\cos\beta)\right]\right\}/a_{32}^6$
		ε	yz	$\varkappa_{32}^2\left\{\dfrac{16}{3} + a_{32}^2 + a_{32}^3 + a_{32}^4 + a_{32}^5 + \dfrac{4}{21}(10 - 3a_{32})\,[2P_2(\cos\beta) - \cos 2\gamma \cdot P_2^2(\cos\beta)]\right. +$ $\left. + \dfrac{16}{7}(a_{32} - 1)\left[4P_4(\cos\beta) + \dfrac{1}{3}\cos 2\gamma \cdot P_4^2(\cos\beta)\right]\right\}/a_{32}^6$
			xz	$\varkappa_{32}^2\left\{\dfrac{16}{3} + a_{32}^2 + a_{32}^3 + a_{32}^4 + a_{32}^5 + \dfrac{4}{21}(10 - 3a_{32})\,[2P_2(\cos\beta) + \cos 2\gamma \cdot P_2^2(\cos\beta)]\right. +$ $\left. + \dfrac{16}{7}(a_{32} - 1)\left[4P_4(\cos\beta) - \dfrac{1}{3}\cos 2\gamma \cdot P_4^2(\cos\beta)\right]\right\}/a_{32}^6$

TABLE 3. Values of C_{hkl} and σ

hkl	C_{hkl} state			σ	
	x	y	z	bcc	fcc
110	−1	−1	2	2	0
101	−1	2	−1	2	0
011	2	−1	−1	2	0
111	0	0	0	0	4
200	−4	2	2	2	4
020	2	−4	2	2	4
002	2	2	−4	2	4
211	−2	1	1	2	0
121	1	−2	1	2	0
112	1	1	−2	2	0
310	−17/5	7/5	2	2	0
130	7/5	−17/5	2	2	0
301	−17/5	2	7/5	2	0
031	2	−17/5	7/5	2	0
103	7/5	2	−17/5	2	0
013	2	7/5	−17/5	2	0
311	−32/11	16/11	16/11	0	4
131	16/11	−32/11	16/11	0	4
113	16/11	16/11	−32/11	0	4
411	−10/3	5/3	5/3	2	0
141	5/3	−10/3	5/3	2	0
114	5/3	5/3	−10/3	2	0

The structure amplitude F_{hkl} is given by

$$F_{hkl} = \sum_p f_p(s) \cdot \exp[2\pi i (\mathbf{r}_p \mathbf{H})]. \tag{14}$$

The summation in Eq. (14) is carried out over atoms in one unit cell. If the substance in question consists of atoms of one kind, Eq. (14) can be simplified by means of Eq. (1):

$$F_{hkl} = \sigma \cdot \sum_k N_k f_k(s), \tag{15}$$

where

$$\sigma = \sum_p \exp[2\pi i (\mathbf{r}_p \mathbf{H})]. \tag{16}$$

In investigations of the anisotropy of the x-ray scattering, the form factor and the structure amplitude can conveniently be separated into a spherically symmetrical part and a part in which the angular dependence of the scattering is allowed for:

$$F_{hkl} = F_0 + \Delta F_{hkl} = \sigma f_0 + \sigma \cdot \Delta f(s), \tag{17}$$

where

$$f_0 = \frac{1}{4\pi} \int f(s) \cdot \sin \beta \, d\beta \, d\gamma.$$

If the valence states are the 2p or 3p shells containing one electron each, the expression for ΔF_{hkl} becomes

$$\Delta F_{hkl}^{(2p)} = 4\pi^2 \sigma C_{hkl} \frac{\varkappa_{2p}^2 d_{hkl}^6}{(d_{hkl}^2 + 4\pi^2 \varkappa_{2p}^2)^4}, \tag{18}$$

$$\Delta F_{hkl}^{(3p)} = \frac{16}{15} \pi^2 \sigma C_{hkl} \frac{\varkappa_{3p}^2 (7 d_{hkl}^2 - 12\pi^2 \varkappa_{3p}^2) d_{hkl}^8}{(d_{hkl}^2 + 4\pi^2 \varkappa_{3p}^2)^6}, \tag{19}$$

where $d_{hkl} = a/\sqrt{h^2 + k^2 + l^2}$; C_{hkl} are the coefficients whose values are given in Table 3.

TABLE 4. Expressions for $\Delta f(s)$ for 3d States

hkl	$\Delta f(s)$					σ	
	z^2-r^2	x^2-y^2	xy	yz	xz	bcc	fcc
110	$\dfrac{4b_H(2+3b_H)}{3(1+b_H)^6}$	$-\dfrac{8b_H(1+3b_H)}{3(1+b_H)^6}$	$-\dfrac{4b_H(2-9b_H)}{3(1+b_H)^6}$	$\dfrac{4b_H(1-3b_H)}{3(1+b_H)^6}$	$\dfrac{4b_H(1-3b_H)}{3(1+b_H)^6}$	2	0
101	$-\dfrac{b_H(4+15b_H)}{3(1+b_H)^6}$	$\dfrac{b_H(4+3b_H)}{3(1+b_H)^6}$	$\dfrac{4b_H(1-3b_H)}{3(1+b_H)^6}$	"	$-\dfrac{4b_H(2-9b_H)}{3(1+b_H)^6}$	2	0
011	"	"	"	$-\dfrac{4b_H(2-9b_H)}{3(1+b_H)^6}$	$\dfrac{4b_H(1-3b_H)}{3(1+b_H)^6}$	2	0
Δf_{av}	$-\dfrac{2b_H^2}{(1+b_H)^6}$	$-\dfrac{2b_H^2}{(1+b_H)^6}$	$\dfrac{4b_H^2}{3(1+b_H)^6}$	$\dfrac{4b_H^2}{3(1+b_H)^6}$	$\dfrac{4b_H^2}{3(1+b_H)^6}$	2	0
111	$-\dfrac{16b_H^2}{3(1+b_H)^6}$	$-\dfrac{16b_H^2}{3(1+b_H)^6}$	$\dfrac{32b_H^2}{9(1+b_H)^6}$	$\dfrac{32b_H^2}{9(1+b_H)^6}$	$\dfrac{32b_H^2}{9(1+b_H)^6}$	0	4
200	$\dfrac{4b_H(2+3b_H)}{3(1+b_H)^6}$	$\dfrac{4b_H(2-9b_H)}{3(1+b_H)^6}$	$-\dfrac{8b_H(1+3b_H)}{3(1+b_H)^6}$	$\dfrac{16b_H}{3(1+b_H)^6}$	$-\dfrac{8b_H(1+3b_H)}{3(1+b_H)^6}$	2	4
020	"	"	"	$-\dfrac{8b_H(1+3b_H)}{3(1+b_H)^6}$	$\dfrac{16b_H}{3(1+b_H)^6}$	2	4
002	$-\dfrac{16b_H(1-3b_H)}{3(1+b_H)^6}$	$\dfrac{16b_H}{3(1+b_H)^6}$	$\dfrac{16b_H}{3(1+b_H)^6}$	"	$-\dfrac{8b_H(1+3b_H)}{3(1+b_H)^6}$	2	4
Δf_{av}	$\dfrac{8b_H^2}{(1+b_H)^6}$	$\dfrac{8b_H^2}{(1+b_H)^6}$	$-\dfrac{16b_H^2}{3(1+b_H)^6}$	$-\dfrac{16b_H^2}{3(1+b_H)^6}$	$-\dfrac{16b_H^2}{3(1+b_H)^6}$	2	4
211	$\dfrac{b_H(4-7b_H)}{3(1+b_H)^6}$	$-\dfrac{b_H(4+5b_H)}{3(1+b_H)^6}$	$-\dfrac{4b_H(3-5b_H)}{9(1+b_H)^6}$	$\dfrac{4b_H(6-b_H)}{9(1+b_H)^6}$	$-\dfrac{4b_H(3-5b_H)}{9(1+b_H)^6}$	2	0
121	"	"	"	$-\dfrac{4b_H(3-5b_H)}{9(1+b_H)^6}$	$\dfrac{4b_H(6-b_H)}{9(1+b_H)^6}$	2	0
112	$-\dfrac{4b_H(2+b_H)}{3(1+b_H)^6}$	$\dfrac{8b_H(1-b_H)}{3(1+b_H)^6}$	$\dfrac{4b_H(6-b_H)}{9(1+b_H)^6}$	"	$-\dfrac{4b_H(3-5b_H)}{9(1+b_H)^6}$	2	0
Δf_{av}	$-\dfrac{2b_H^2}{(1+b_H)^6}$	$-\dfrac{2b_H^2}{(1+b_H)^6}$	$\dfrac{4b_H^2}{3(1+b_H)^6}$	$\dfrac{4b_H^2}{3(1+b_H)^6}$	$\dfrac{4b_H^2}{3(1+b_H)^6}$	2	0

It must be stressed that the values of ΔF_{hkl} are independent of the signs of the reflection indices. Moreover, the expression for ΔF_{hkl} corresponding to the reflections nh, nk, and nl is of the same form as for the reflections hk$l$. The only difference is in the value of the interplanar distance d.

The values of $\Delta f(s)$ for the 3d states are given in Table 4. This table includes also the average values Δf_{av} which correspond to the changes in the structure amplitude resulting from the anisotropy of the x-ray scattering in polycrystalline samples. The following notation is used in Table 4:

$$b_H = \varkappa_{3d}^2 s^2 = \frac{4\pi^2 \varkappa_{3d}^2}{a^2} (h^2 + l^2 + k^2).$$

In the case of electron scattering, the changes in the structure amplitudes $\Delta\Phi_{hkl}$ are given by

$$\Delta\Phi_{hkl} = - K \frac{\Delta F_{hkl}}{s^2} .$$

Thus, the changes in the structure amplitude resulting from the anisotropy of electron scattering in the case of the 2p, 3p, and 3d states can be found by multiplying Eqs. (18) and (19) or the formulas in Table 4 by the factor $-K/s^2$.

Discussion of Results

An analysis of Eqs. (18) and (19) and of the formulas in Table 4 shows that $\Delta F_{av} = 0$ for the p states. This means that there is no anisotropy in the scattering of x rays or electrons by a polycrystalline sample.

The scattering anisotropy should be absent also in the case of diffraction by single crystals if the valence electrons "resonate" between three p states. However, if an electron is in some fixed p state, a study of the scattering of x rays or electrons should reveal effects resulting from the anisotropy in the spatial distribution of the atomic charges. The changes in the intensities in the x-ray diffractograms, compared with the case of a spherically symmetrical distribution of the electron density, may reach 20–40% for elements in the second period and 7–15% for elements in the third period.

In the case of transition metals, which have partly filled 3d shells, the anisotropy in the scattering of x rays and electrons should be observed irrespective of whether a sample is a single crystal or a polycrystalline aggregate.

The changes in the relative intensity of the scattering of x rays by polycrystalline samples, compared with the spherically symmetrical case, should be greatest between Sc ($s = 6.4$ Å^{-1}) and Ni ($s = 11.3$ Å^{-1}).

For these values of s, the change in the relative intensity of diffraction by transition metals should be 4–5% .

In the electron diffraction case, the greatest changes in the relative intensities of the scattering by single crystals of transition metals should be observed for the first reflections.

If electrons are scattered from polycrystalline samples, the effects in question should be greatest for $s = 3.3$–7.6 Å^{-1} and the changes in the relative intensities should then amount to about 1%.

Literature Cited

1. J. C. Slater, Phys. Rev., 36:57 (1930).
2. F. C. von der Lage and H. A. Bethe, Phys. Rev., 71:612 (1947).
3. B. K. Vainshtein, Structure Analysis by Electron Diffraction, Pergamon, Oxford (1964).

RELATIONSHIP BETWEEN THE REFLECTION OF
X RAYS AND THE DEGREE OF PERFECTION OF
CRYSTALS IN FOCUSING MONOCHROMATORS*

N. M. Olekhnovich

The results are described of a study of the relationship between the reflection intensity of x rays and the degree of perfection of crystals used in focusing monochromators. An expression has been obtained which gives the variation in the intensity with the integrated reflection width and the size of the focus of the radiation source. Analytical conditions are presented for estimating the optimal mosaic parameters (size and disorientation angle of the blocks), which the monochromator crystal should satisfy to obtain the maximum reflection intensity.

The use of monochromatic radiation is required for the successful solution of many problems associated with the study of the intensities of coherent and incoherent scattering of x rays by crystalline and amorphous solids. Consequently, the construction of strong monochromatic sources merits great attention. The use of focusing monochromators [1—3] leads to an appreciable gain in intensity compared with plane monochromators. For focusing monochromators, the intensity of the reflected beam, its spectral composition, and its geometry depend essentially on the size and brightness of the source (the focus of an x-ray tube), the perfection of the crystal used as the monochromator, and the focusing conditions [4].

The effect of the degree of perfection of monochromator crystals on the basic characteristics of focusing monochromators, and particularly on the intensity of the reflected beam, has not received its due attention in the literature. For example, it has been considered [5] that tubes with small focus dimensions and the most perfect crystals must be used to obtain a narrow intense beam of monochromatic radiation. However, measurements made on quartz of different degrees of perfection [6] showed that the maximum intensity of a monochromatic beam is observed for "average" degrees of perfection of the monochromator crystal.

In the present investigation, the problem was posed of analyzing how the degree of perfection of a crystal affects the total reflection intensity of a monochromatic beam in focusing monochromators, of estimating the optimum mosaic parameters (size and angle of disorientation of the blocks) of the crystal at which maximum reflection intensity is obtained, and how these parameters vary with the size of the focus of the x-ray tube and the geometry of the focusing monochromator.

When discussing the conditions for focusing an x-ray beam in focusing monochromators,

---

* "Semiconductors," pp. 116–122 (see page 3).

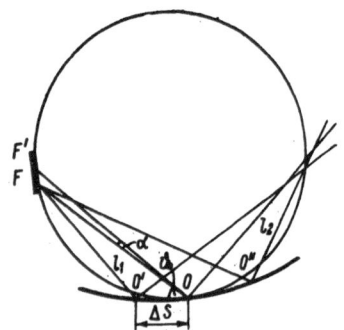

Fig. 1. Focusing by a monochromator crystal (Johann).

it is usually assumed that the focus of the x-ray
tube is a point. However, even in the case of
fine-focus tubes, the dimensions of the focus
cannot be neglected, because its angular width,
the angle subtended by the focus at the center of
the monochromator (Fig. 1), may be several
minutes of arc, which exceeds appreciably the
angular width of the reflection from a perfect
crystal. Let the ray leaving the point F of the
tube focus impinge on a focusing monochromator
crystal at a Bragg angle ϑ_0 at a point O. Then,
the ray leaving the point F' will impinge at the
point O' at an angle of $\vartheta_0 + \alpha$.

The reflection intensity is zero for rays for which the angle α is higher in absolute value
than the integrated width of the reflection from the monochromator crystal. The effective width
of the focus area whose intensity will make a contribution to the reflected beam will evidently
decrease with decreasing integrated width of the reflection curve, i.e., the more perfect the
monochromator crystal. Consequently, to obtain the maximum intensity for the reflected monochromatic beam, it is necessary to establish a definite optimum relationship between the effective width of the focus and the integrated width of the reflection curve of the monochromator
crystal.

Let the intensity distribution at the focus be represented by the function $G(\alpha)$, the reflectivity of the monochromator crystal by $R(\vartheta, \alpha, \lambda)$, and the spectral intensity distribution of the
primary beam by $i(\lambda - \lambda_0)$. The reflection intensity from the monochromator is proportional
to the product of these functions [6]. The total intensity of the monochromator beam when the
monochromator is placed at the Bragg angle $(\vartheta = \vartheta_0)$ is then defined by the following expression:

$$I_m = \int_{-\Delta S_0}^{\Delta S_0} \int_{-\infty}^{\infty} \int_{-\infty}^{\infty} i(\lambda - \lambda_0) G(\alpha) R\left[\alpha + (\lambda - \lambda_0) \frac{d\vartheta}{d\lambda} + \eta\left(\frac{\Delta S \sin \vartheta_0}{l_1}\right)\right] \frac{dS \sin \vartheta_0}{l_1} \, d\lambda \, d\alpha, \qquad (1)$$

where

$$(\lambda - \lambda_0) \frac{d\vartheta}{d\lambda} = \frac{\lambda - \lambda_0}{\lambda} \tan \vartheta_0$$

is the dispersion (the change in the Bragg angle for radiation having a wavelength λ relative to
radiation having a wavelength λ_0), and η is a term which takes account of the change in the
angle of incidence as a function of the position of the reflecting point on the surface of the monochromator with respect to its center O.

For simplicity, let us restrict ourselves to the case where the divergence of the incident
beam is low and the term η can be neglected. The intensity of the reflected beam will then be
proportional to the width of the illuminated surface of the monochromator (or to the angle of
divergence of the incident beam $\Delta \varphi_0$). With this condition, Eq. (1) becomes

$$I_m = \Delta\varphi_0 \int_{-\infty}^{\infty} i(\lambda - \lambda_0) \int_{-\infty}^{\infty} G(\alpha) R\left[\alpha + (\lambda - \lambda_0) \frac{d\vartheta}{d\lambda}\right] d\lambda \, d\alpha. \qquad (2)$$

The dependence of the change in I_m on the focus parameters of the x-ray tube and the degree of perfection of the monochromator crystal can be traced for a specific model without affecting the generality of the qualitative conclusions. Let the intensity distribution at the focus be represented by a Gaussian function $G(a) = G_0 \exp[-\pi a^2/w_g^2]$. Here, the total power of the radiation source, $P = \int G(a)da = G_0 w_g$, where $w_g = d_0/l_1$, is the effective angular width of the focus (d_0 being the linear width of the focus). Let us assume that the reflection curve for the monochromator crystal is also represented by a Gaussian function

$$R = R_0 \exp \left[- \frac{\pi \left(a + (\lambda - \lambda_0) \dfrac{d\vartheta}{d\lambda} \right)^2}{w_r^2} \right].$$

The integrated reflection is then $\rho = \int R(a)\, da = R_0 w_r$, while the integrated width of the reflection curve is w_r. Since we are interested only in the intensity of monochromatic K_α radiation, we shall restrict ourselves to considering the K_{α_1} line in the radiation spectrum from the focus, in which the intensity distribution is represented by the function

$$i(\lambda - \lambda_0) = i_\nu \left[- \frac{\pi (\lambda - \lambda_0)^2}{w_j^2} \right]$$

with a total intensity $I_0 = \int i(\lambda - \lambda_0)\, d\lambda = i_0 w_j$ and a width w_j.

Putting the values of these functions in Eq. (2) and integrating, we obtain the following expression for the total intensity of a monochromatic beam:

$$I_m = \Delta \varphi_0 \frac{i_0 w_j G_0 w_g R_0 w_r}{\sqrt{w_r^2 + w_g^2 + w_j^2 \left(\dfrac{d\vartheta}{d\lambda} \right)^2}} = I_0' \frac{\rho}{\sqrt{w_r^2 + w_g'^2}} , \qquad (3)$$

where $I_0' = \Delta \varphi_0 I_0 P$ is the intensity incident on the surface of the monochromator crystal;

$$w_g'^2 = w_g^2 + w_j^2 \left(\frac{d\vartheta}{d\lambda} \right)^2.$$

From Eq. (3), we see that the intensity of the monochromatic beam is determined only by the magnitude of the integrated reflection and the integral width w_r. Furthermore, it also follows from Eq. (3) that sharp focusing for constant radiation power gives more intense monochromatic beams.

To find the condition for obtaining maximum intensity I_m, we must know how the integrated reflection of the crystal ρ is related to w_r. This relationship may be obtained approximately from the extinction theory of crystals.

From the model for taking account of the secondary extinction of mosaic crystals [7], it can be established that the reflection intensity from a crystal exhibiting both primary and secondary extinctions is determined by the following relationship:

$$\frac{I}{I_0} = R(\xi) = \left\{ Q'B \frac{L}{\sin \vartheta_0} \exp[-\pi (B\xi)^2] \right\} \times$$

$$\times \left\{ 1 - \left\{ 1 - Q'B \frac{L}{\sin \vartheta_0} \exp[-\pi (B\xi)^2] \right\}^2 \exp[-2\mu L/\sin \vartheta_0] \right\}^{-1}, \qquad (4)$$

where

$$B = \frac{2L \cos \vartheta_0/\lambda_0}{\sqrt{1 + 8\pi\overline{\varphi^2}L^2 \cos^2 \vartheta_0/\lambda^2}};$$

L is the average size of the mosaic blocks; $\sqrt{2\pi\overline{\varphi^2}}$ is the average angle of disorientation of the blocks; $Q' = fQ$ is the reflectivity of the crystal when account is taken of primary extinction.

Expanding the denominator of Eq. (4) in series and taking only the first-order small terms, we find that the reflection intensity at the maximum in the reflection curve ($\xi = 0$) is determined by the relationship

$$R_0 = Q'B/2(\mu + Q'B). \qquad (5)$$

Integration of Eq. (4) in accordance with [7] gives an expression for the integrated reflection intensity which may be written to a first approximation in the form

$$\rho = Q'/2(\mu + gQ'B), \qquad (6)$$

where $g = 1/\sqrt{2}$.

Since the integrated width of the reflection curve is determined by the ratio ρ/R_0, we may write

$$w_r = \frac{1}{B} \cdot \frac{1 + Q'B/\mu}{1 + gQ'B/\mu} \approx k/B, \qquad (7)$$

where

$$k \approx \frac{1 + Q'/\mu w_r}{1 + gQ'/\mu w_r}$$

changes very little with an increase in B. Inserting the obtained values of ρ in Eq. (3), having regard to Eq. (7), we obtain

$$I_m = I_0' \frac{Q'}{2\mu\left(1 + \frac{g'Q'}{\mu w_r}\right)\sqrt{w_r^2 + w_g'^2}} = I_0'\left[2g'\left(\frac{w_g'}{g'Q'/\mu} + \frac{w_g'}{w_r}\right)\sqrt{1 + \frac{w_r^2}{w_g'^2}}\right]^{-1}, \qquad (8)$$

where $g' = kg$.

Figure 2 shows the relationship between I_m and the integrated width of the reflection curve w_r for $w_g' = 3 \cdot 10^{-3}$ rad, which represents about 0.2 mm of the linear width of the tube focus for $l_1 = 70$ mm. This corresponds approximately to the geometrical operating conditions of focusing monochromators in normal x-ray diffractometers. The quantity $g'Q'/\mu$ was assumed to be $2 \cdot 10^{-3}$ (curve 1), the value corresponding to LiF (200 reflection), and $1 \cdot 10^{-3}$ (curve 2), corresponding to Ge (111) for Cu K_α radiation.

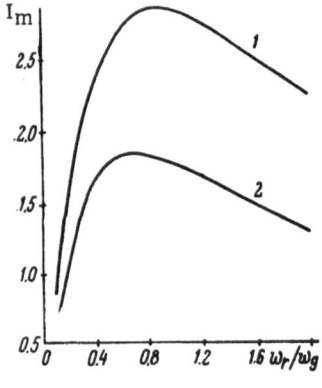

Fig. 2. Relationship between the intensity (in relative units) of a monochromatic beam and the integrated reflection width of a monochromator crystal.

As may be seen from Fig. 2, the intensity function of a beam reflected from the focusing monochromator passes through a maximum, whose value increases with the reflectivity of the crystal. With a decrease in reflectivity of the monochromator crystal, the maximum is displaced toward lower values of w_r.

At lower values of w_r, the intensity curve falls rapidly from the maximum, reaching relatively low values. This leads to the conclusion that the use of extremely perfect crystals for focusing monochromators at $w_g^! = 3 \cdot 10^{-3}$ is disadvantageous for obtaining the maximum intensity of the monochromatic beam. An analysis of the curves in Fig. 2 also shows that crystals having an integrated width lying in the range $\pm 0.3 - 0.4$ of the value at which the maximum I_m is reached may be used for focusing monochromators without substantial losses in intensity.

From a study at the extremum of Eq. (8), it is seen that I_m reaches a maximum value (I_{max}) at an integrated width defined by the following relationship:

$$(w_r)_m = (g'Q'/\mu)^{1/3} w_g^{'\,2/3}. \tag{9}$$

Here,

$$I_{max} = I_0' \frac{Q'}{2\mu\,[(g'Q'/\mu)^{2/3} + (w_g')^{2/3}]^{3/2}} = \frac{I_0'}{2\mu w_g'[(1/Q')^{2/3} + (g'/\mu w_g')^{2/3}]^{3/2}}. \tag{10}$$

Equation (9) shows that, if $g'Q'/\mu < w_g^!$, then the maximum intensity of a monochromatic beam will be observed for $(w_r)_m < w_g^!$ and, on the other hand, if $g'Q'/\mu > w_g^!$, then $(w_r)_m > w_g^!$.

The maximum reflection intensity for a given focus width will be higher, the greater the reflectivity of the crystal. This means that Q' should be equal to Q, the reflectivity of an ideally mosaic crystal. Consequently, to obtain the greatest intensity I_m, crystals must be selected which have no appreciable primary extinction, i.e., the size of the mosaic blocks should not exceed 10^{-4} cm. Crystals with relatively high dislocation densities satisfy this condition.

The optimum angle of disorientation between the mosaic blocks may be estimated from the optimum integrated width of the reflection curve. Putting the values of B and $(w_r)_m$ in Eq. (7) and solving the equation obtained for $2\pi\overline{\varphi^2}$, we obtain

$$\sqrt{2\pi\overline{\varphi^2}} = \frac{1}{2}\left[k^{-4/3}\left(\frac{gQw_g^{'2}}{\mu}\right)^{2/3} - (\lambda/2L_m\cos\vartheta_0)^2\right]^{1/2} \tag{11}$$

The range of optimum values for the average angle of disorientation of the blocks at $w_g^! = 3 \cdot 10^{-3}$ is $2-6'$. However, if it is necessary to obtain narrower monochromatic beams, crystals having the lowest angle of disorientation within this range must be chosen.

The results obtained may be used as a basis for making the correct choice of crystals with respect to degree of perfection in making focusing monochromators and in choosing the method of preparing them.

I am deeply grateful to Academician N. N. Sirota of the Belorussian Academy of Sciences, for valuable advice and great interest in the investigation.

Literature Cited

1. H. H. Johann, Z. Phys., 69:185 (1931).
2. T. Johansson, Z. Phys., 82:507 (1933).
3. B. Ya. Pines, in: Symposium for the 70th Birthday of Academician A. F. Ioffe [in Russian], Izd. Akad. Nauk SSSR, Moscow (1950), p. 448.
4. A. Guinier, Théorie et technique de la radiocristallographie, Dunod, Paris (1956).
5. A. Guinier, Ann. Phys. (Paris), 12:161 (1939).
6. J. Čermák, Czech. J. Phys., B12:602 (1962).
7. A. V. Kuznetsov, Kristallografiya, 7:121 (1962).

ELECTRON DENSITY DISTRIBUTION AND
EFFECTIVE CHARGES IN INDIUM PHOSPHIDE*

N. N. Sirota, E. M. Gololobov, and A. U. Sheleg

The experimental values of the structure amplitudes, determined in earlier investigations, were used to find the electron density distribution in the indium phosphide lattice. Electron density "bridges," whose density was at least 0.35 ± 0.05 electron/$\overset{\bullet}{A}{}^3$, were found between the nearest unlike atoms. The signs and the effective charges of the atoms (0.40 ± 0.15 el) were determined.

Studies of the electron density distribution provide a direct method for the experimental and the theoretical elucidation of the nature of the chemical bonds between atoms in crystals.

We used the x-ray diffraction method to determine the distribution of the electron density in indium phosphide from the structure factors obtained in an earlier experimental study [1].

The structure factors of InP were obtained from the x-ray diffraction spectra of powder samples, 20 mm in diameter and 2–3 mm thick. These samples consisted of finely ground and elutriated powders whose grain size was below 1μ. These powders were compacted at pressures up to 6000 kgf/cm^2. Some of the samples were not compacted: in these cases a binder was used. The samples containing a binder were ground and polished. The surfaces of the compacted samples had a sufficiently fine finish because we used a plunger with a mirror-smooth surface.

The integrated intensities of the reflections were measured by means of a URS–50IM diffractometer, an NaI:Tl scintillation counter, and an amplitude analyzer. We used Cu K$_\alpha$ radiation made monochromatic by means of a bent single-crystal germanium plate cut along the (111) plane.

The influence of porosity had to be eliminated in the precise measurements of the x-ray reflections from the powders. This was usually achieved by increasing the compacting pressure during the preparation of the samples. However, a preferential orientation of some crystallographic planes could appear in strongly compacted samples. Our investigation of the effects of porosity and of the preferential orientation established that the intensities of the 220 and 440 reflections increased strongly when the compacting pressure was increased (Fig. 1). A comparison of the ratios of the intensities of various reflection pairs obtained for compacted samples with the corresponding ratios for samples prepared without compression indicated that the intensities of the 220 and 440 reflections were affected by the preferential orientation of the (110) crystallographic plane. This plane was in the direction of easiest cleavage in $A^{III}B^V$ compounds

---

* "Semiconductors," pp. 123–127 (see page 3).

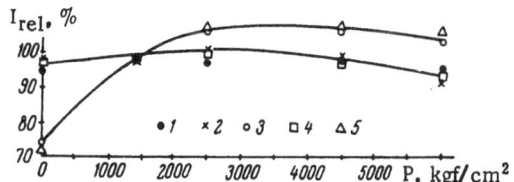

Fig. 1. Dependences of the relative intensities of the reflections I_{rel} on the compacting pressure P applied to InP: 1) 111; 2) 200; 3) 220; 4) 311; 5) 440.

[2] because of the ionic-covalent nature of the chemical bonds in these compounds. Therefore, when the powders were ground, this plane emerged preferentially on the particle surfaces. In the samples compacted at high pressures, the (110) plane was oriented in a certain manner relative to the surface of the sample. A similar effect was reported in [3] for the (100) plane in vanadium nitride.

The structure factors of InP were determined using samples compacted by pressures of 2500 kfg/cm² but the intensities of the 220 and 440 reflections were corrected for the preferential orientation. These corrections were found by comparing the relative intensities obtained for samples which were not compressed with the intensities found for the compacted samples. Corrections were also made for the thermal diffuse scattering [4]. The extinction effects were slight because we used very fine powders. The linear absorption coefficient μ was assumed to be 993 cm^{-1} [5]. The polarization factor was found from the expression [6]

$$p = \frac{1 + \gamma \cos^2 2\vartheta_M \cdot \cos^2 2\vartheta}{1 + \gamma \cos^2 2\vartheta_M},$$

where $\gamma = 1.17 \pm 0.01$ [1].

The relative intensities were converted to the absolute values by measuring the intensity of the incident beam or by comparing the results with those obtained for a nickel standard. The difference between these results amounted to about 1%, i.e., it was within the limits of the experimental error.

The temperature factors $M = B \sin^2 \vartheta / \lambda^2$ for different types of atom were determined from the condition for a minimum value of

$$\sum_{hkl} \frac{f_{\exp}^{(hkl)} e^{M} - f_{\text{theor}}^{(hkl)}}{f_{\text{theor}}^{(hkl)}}$$

for 12 reflections obtained at large reflection angles. Here, $f_{\text{theor}} = f_0 + \Delta f'$, the dispersion corrections $\Delta f'$ being taken from [7]. Our values $B_{In} = 1.5$ Å² and $B_P = 1.3$ Å² were in agreement with the values $B_{In} = 1.69$ Å² and $B_P = 1.34$ Å² deduced from the x-ray determinations of Θ_D at various temperatures [8].

Figure 2 gives the theoretical values of the atomic scattering factors of indium and phosphorus calculated by the Hartree–Fock–Slater method [9], as well as the experimental values of these factors [1].

The electron density distribution was calculated using the three-dimensional Fourier difference series

$$\rho(x, y, z) = \frac{1}{V} \sum_{h} \sum_{k} \sum_{l} F_{3,hkl} \exp\left[-2\pi i (hx + ky + lz)\right] + A_1 e^{-\alpha_1 r^2} + A_2 e^{-\alpha_2 r^2},$$

where

$$F_3 = F_{\exp} - F_1 - F_2.$$

The experimental f curves, obtained at 20°C, were extrapolated to higher values of

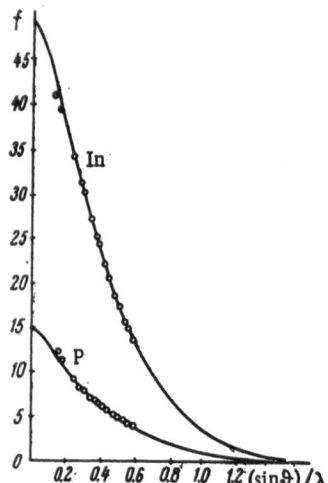

Fig. 2. Values of the atomic scattering factors of indium and phosphorus (the circles represent the experimental values for InP and the continuous curves are the theoretical dependences found by the Hartree–Fock–Slater method [9] for free atoms).

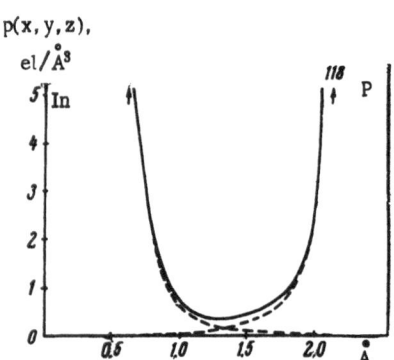

Fig. 3. Electron density distribution between the phosphorus and the indium atoms along the [111] direction in InP.

$(\sin\vartheta)/\lambda$ employing the theoretical values found by the Hartree–Fock–Slater method (Fig. 2). Next, the f curves for $(\sin\vartheta)/\lambda > 0.5$ were approximated by two Gaussian functions from which the coefficients A_1 and A_2, α_1 and α_2 were determined.

Figure 3 shows the electron density distribution between the nearest In and P atoms along the [111] direction in the (110) plane of InP. The dashed curves represent the electron densities on the opposite sides of the In and the P atoms. It is evident from Fig. 3 that the electron density between the nearest indium and phosphorus atoms in InP fell to a minimum value of 0.35 ± 0.05 electron/Å^3, i.e., an "electron density bridge" typical of covalent bonds was observed between these atoms. An analysis of our structure factors $|F|^2$, particularly of the difference-type structure amplitudes corresponding to small values of $(\sin\vartheta)/\lambda$, which were most sensitive to the redistribution of electrons between atoms in compounds, indicated an appreciable ionicity

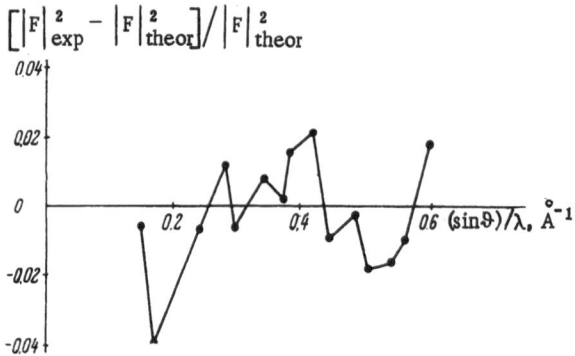

Fig. 4. Relative deviations of the experimental structure factors from theoretical values found for different reflections in InP.

of the bonding in InP. Figure 4 shows relative the deviations of the experimental structure factors from the theoretical values found from the Hartree–Fock–Slater f factors of neutral atoms. The greatest deviation (4%), with the experimental values smaller than the theoretical factor, was observed for the 200 reflection. This indicated that the electrons in InP were transferred from the indium to the phosphorus atoms. Our estimates of the effective charges of the atoms, obtained by the method we proposed in [10], gave 0.40 ± 0.15 el ("el" here and in Fig. 3 denotes "electron").

We thus established a considerable ionicity of the bonding in the predominantly covalent indium phosphide.

Literature Cited

1. N. N. Sirota and E. M. Gololobov, Vestsi Akad. Navuk Belarus. SSR, Ser. Fiz.-Mat. Navuk, No. 5, p. 92 (1969).
2. H. Pfister, Z. Naturforsch., 10a:79 (1955).
3. S. Hosoya, T. Yamagishi, and M. Tokonami, Tech. Rep. ISSP, Ser. A, No. 275 (1967); J. Phys. Soc. Jap., 24:368 (1968).
4. D. R. Chipman and A. Paskin, J. Appl. Phys., 30:1992 (1959).
5. International Tables for X-Ray Crystallography, Vol. 3, Kynoch Press, Birmingham (1962).
6. N. M. Olekhnovich, Kristallografiya, 14:261 (1969).
7. D. T. Cromer, Acta Crystallogr., 18:17 (1965).
8. V. V. Rozov and N. N. Sirota, in: Chemical Bonds in Semiconductors and Solids (ed. by N. N. Sirota), Consultants Bureau, New York (1967), p. 143.
9. H. P. Hanson, F. Herman, J. D. Lea, and S. Skillman, Acta Crystallogr., 17:1040 (1964).
10. N. N. Sirota and E. M. Gololobov, Dokl. Akad. Nauk SSSR, 156:1075 (1964).

DISTRIBUTION OF THE UNCOMPENSATED ELECTRON DENSITY IN THE FERROMAGNETIC MANGANESE COMPOUNDS MnAs, MnSb, and MnBi*

N. N. Sirota and É. A. Vasil'ev

Values of the magnetic form factors of manganese atoms were obtained from data on the intensity of magnetic neutron scattering in ferromagnetic manganese compounds. The distribution of the density of uncompensated electrons, obtained from the Fourier transform of the magnetic form factor curves, gives an indication of the nature of the change in the amount of overlap of 3d electron shells and of the connection between physical properties and electron distribution.

The intermetallic compounds MnAs, MnSb, and MnBi crystallize with a nickel arsenide-type structure, and they have many features in common regarding their structure and nature of physical properties. At room temperature, all these compounds are ferromagnetic. The Curie point of manganese arsenide is 45°C; of MnSb and MnBi it is 314 and 360°C, respectively. Their ferromagnetism is due to the uncompensated spins of the 3d electron shells of the manganese atoms, located at the crystal lattice points with the coordinates 0, 0, 0 and 0, 0, $\frac{1}{2}$.

The characteristics of magnetic materials which are important for an understanding of the nature of the magnetic state of matter can be obtained from an investigation of neutron scattering by magnetic moments in ferro- and antiferromagnetic materials.

The position, magnitude, and orientation of magnetic moments in a crystal lattice can, for example, be determined by neutron diffraction. It is of particular interest to be able to study the distribution of the density of uncompensated electrons causing magnetic ordering.

In order to do this, the angular dependence of the intensity of magnetic neutron scattering was determined in ferromagnetic binary manganese compounds. Neutron diffraction patterns of MnAs, MnSb, and MnBi powders were obtained above the Curie point of the alloys and at temperatures in the ferromagnetic ordering region. The neutron diffraction patterns of one of the alloys studied, manganese antimonide, shown in Fig. 1, indicate the nature of the change of intensity of the lines on varying the temperature from 130°K to temperatures above the Curie point.

The nuclear scattering intensities were used as a scale for obtaining the absolute values of the magnetic scattering intensities.

---

* "Semiconductors," pp. 128–134 (see page 3). This article was not presented at the International Symposium on Chemical Bonds in Semiconducting Crystals, Minsk, 1967.

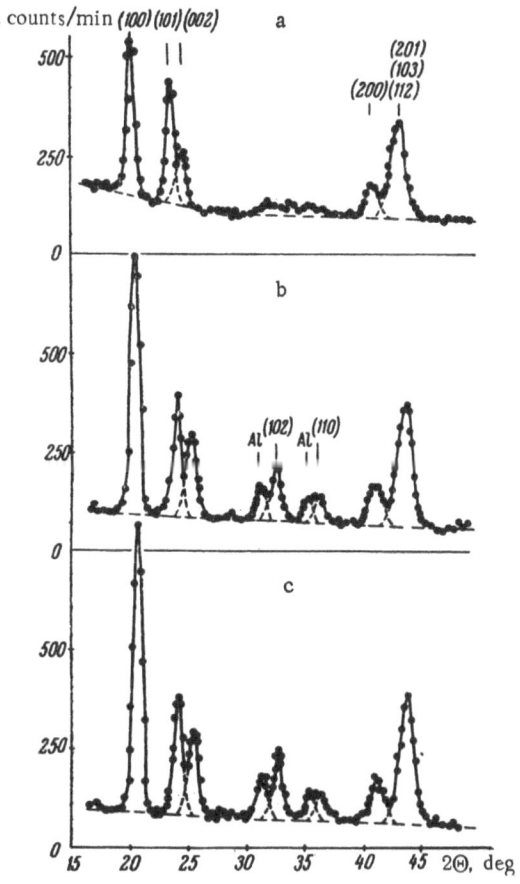

Fig. 1. Neutron diffraction patterns of manganese antimonide in the ferromagnetic ordering region at temperatures: a) above the Curie point (T = 620°K > θ_c); b) T = 293°K < θ_c; c) T = 130°K < θ_c.

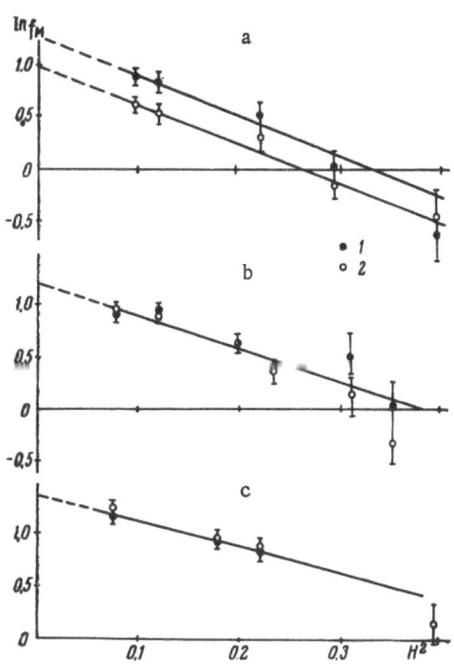

Fig. 2. Dependence of the logarithms of magnetic form factors on H = (2 sin Θ)/λ in ferromagnetic manganese compounds: a) MnAs at T = 130°K, f_M = 3.5 exp (−3.1 H^2) (1), and 293°K, f_M = 2.7 exp (−3.1 H^2) (2); b) MnSb, f_M = 3.3 exp (−2.5 H^2) at T = 130 (1), and 293 (2)°K; c) MnBi, f_M = 3.8 exp (−2.25 H^2). 1) From our data at T = 293°K, 2) from data in [3].

The magnetic scattering intensity is determined principally by the value of the structure amplitude F(khl), which involves the magnetic scattering amplitude p:

$$I_M = Kjq^2 \frac{|F(hkl)|^2}{\sin\Theta \ \sin 2\Theta} \exp(-2W) = Kjq^2 \frac{|\sum_n \exp 2\pi i (hx_n + ky_n + lz_n)|^2}{\sin\Theta \ \sin 2\Theta} \exp(-2W), \qquad (1)$$

where x, y, z are the coordinates of the scattering atom; h, k, l are the indices of the reflecting plane; j is the frequency factor; q^2 is the square of the magnetic interaction vector, taking into account the dependence of intensity on the angle between the direction of the magnetic moment and the scattering vector; exp(− 2W) is the temperature factor. The constant K represents the properties of the sample (size, density) and the instrument parameters.

The magnetic scattering amplitude can be represented in the form

$$p = \left(\frac{e^2\gamma}{mc^2}\right)^2 Sf = 0.539 \cdot 10^{-12} Sf = 0.27 n_B f \cdot 10^{-12} \ cm, \qquad (2)$$

where n_B is the magnetic moment of the atom; f is the form factor normalized to unity; this factor takes into account the angular dependence of the intensity of magnetic neutron scattering.

TABLE 1. Values of the Magnetic Moments of Manganese Atoms at T = 293°K

Compound	n, μ_B (magnetic measurements)	n, μ_B (neutron diffraction data)
MnAs	2, 5	2, 7
MnSb	3, 2	3, 3
MnBi	3, 6	3, 8

The value of $n_B f$, which by analogy with the atomic amplitude of x-ray scattering can be called the atomic amplitude of magnetic scattering, or the magnetic form factor $n_B f = f_M(H)$, can be calculated directly from the intensity of magnetic scattering if the nuclear and magnetic structure of the compound are known. When the value of the reciprocal lattice vector is $H = (2 \sin\Theta)/\lambda = 0$, the magnetic form factor f_{M_0} determines the effective number of electrons taking part in the scattering.

The dependence of f_M on H, as is shown, for example, in [1], can be approximated by a function of the form

$$f_M = \sum_i f_i \exp(-\alpha_i H^2). \qquad (3)$$

The experimental values of the magnetic form factors of the compounds at different temperatures show that satisfactory agreement is obtained between the form factor and the approximating function by taking the first term on the right-hand side of Eq. (3). In this case, by taking logarithms in Eq. (3), we obtain a linear dependence $\ln f_M = f(H^2)$ (Fig. 2). The slope of the straight line obtained gives the value of α, and f_{M_0}, determined by extrapolating the straight line to $H = 0$, gives the value of the magnetic moment.

The agreement, within the experimental error, for values of n_B obtained by direct magnetic measurements and from neutron diffraction data (cf. Table 1), shows that this approximation can be used.

Thus, the magnetic moment can be determined independently by neutron diffraction measurements; this is especially important in studying antiferromagnetics.

The values of α which we obtained were different for all three compounds studied, which indicates some differences in the dependence of the magnetic form factor on the scattering

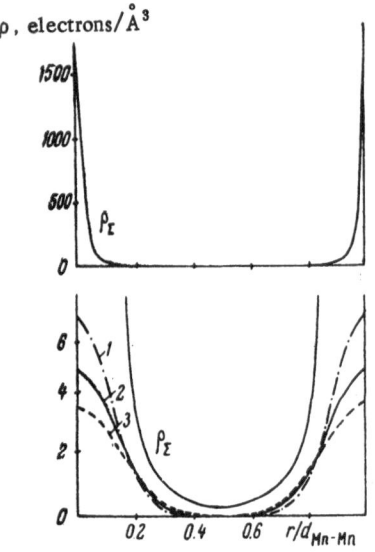

Fig. 3. Distribution of the uncompensated electron density in the [001] direction for MnBi (1), MnSb (2), and MnAs (3). ρ_Σ is the density of all the electrons in the neutral manganese atom, calculated from the theoretical atomic form factor [4].

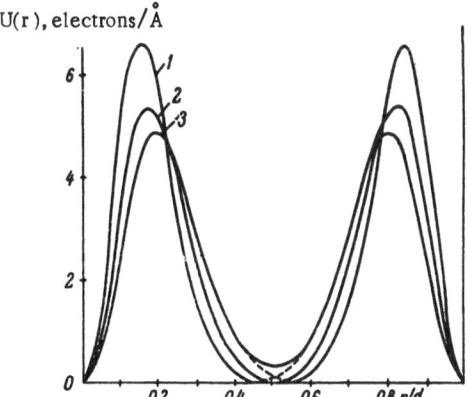

Fig. 4. Radial distribution of the electron density in equiatomic manganese compounds MnBi (1), MnSb (2), and MnAs (3).

angle. These data are supported by results on the magnetic form factors of manganese in MnSb and MnBi in [2] and [3].

The distribution of electron density giving rise to ferromagnetism is described by the Fourier transform of the dependence of the magnetic form factor on the reciprocal lattice vector H.

In the case being considered, when the form factor is approximated by the exponential function (3), the electron density distribution has the form

$$\rho(r) = f_{M_0} \left(\frac{\pi}{\alpha}\right)^{3/2} \exp\left(-\frac{\pi^2}{\alpha} r^2\right). \tag{4}$$

As can be seen in Fig. 3, the distinctive feature of the density distribution in MnAs, MnSb, and MnBi is the slower decrease of ρ in MnAs, when r is increased, than in MnSb and particularly MnBi. As a result, the maximum value of the electron density halfway between the atoms, i.e., where there is a possible overlap of the electron shells of the manganese atoms, is found in manganese arsenide. In absolute value, the electron density at $r/d_{Mn-Mn} = 0.5$ is 0.01 electron/Å^3, or 0.3% of the maximum value corresponding to $r = 0$.

The existence of some overlap of the 3d shells in manganese arsenide is particularly noticeable in curves showing the radial distribution of the electron density $U(r) = 4\pi r^2 \rho(r)$ (see Fig. 4). The maximum of the radial distribution, the position of which may be associated with the radius of the 3d shell, moves to higher values of r on reducing the atomic number of the anion. The following values of r_{max} were obtained for the compounds studied: for MnBi 0.48, MnSb 0.50, and MnAs 0.56 Å.

The increase in the overlap of the 3d shells observed in the series of compounds MnBi — MnSb — MnAs is due both to an increase of the radius of the electron shell and to a decrease of the distance between atoms in the [001] direction, which varies from $d_{Mn-Mn} = 2.85$ Å in manganese arsenide to $d_{Mn-Mn} = 3.06$ Å in MnBi. The overlap of the 3d-shells of manganese atoms, which we noted, agrees well with conclusions in [5] based on measurements of the electrical conductivity of manganese arsenide and antimonide. In addition to this, in the overlap of 3d electron shells in compounds with a nickel arsenide-type structure, there is a direct exchange interaction which, according to data in [6], causes antiferromagnetic ordering. The existence of ferromagnetism in the compounds studied indicates that the overlap of 3d electron shells is not so large as to change significantly the nature of the exchange interaction.

Literature Cited

1.   N. M. Olekhnovich and N. N. Sirota, in: Chemical Bonds in Semiconductors and Solids (ed. by N. N. Sirota), Consultants Bureau, New York (1967), p. 65.
2.   W. J. Takei, D. E. Cox, and G. Shirane, Phys. Rev., 129:2008 (1963).
3.   B. W. Roberts, Phys. Rev., 104:607 (1956).
4.   R. E. Watson and A. J. Freeman, Acta Crystallogr., 14:27 (1961).
5.   G. Fisher and W. B. Pearson, Can. J. Phys., 36:1010 (1958).
6.   J. B. Goodenough, A. Wold, R. J. Arnott, and N. Menyuk, Phys. Rev., 124:373 (1961).

DISTRIBUTION OF THE POTENTIAL
IN THE GALLIUM PHOSPHIDE LATTICE*

N. N. Sirota, A. U. Sheleg, and Zh. A. Matskevich

The x-ray diffraction f curves, determined experimentally, were used in computer calculations of the distribution of the potential in the (110) and (100) planes of gallium phosphide crystals. It was found that the potential between the nearest Ga and P atoms rose to 15.21 V along the [111] direction. The average internal potential in gallium phosphide, deduced from experimental data, was 16.02 V, whereas the value calculated from the diamagnetic susceptibility was 14.5 V.

The distribution of the potential in a crystal lattice, like the electron density distribution, is an important characteristic of matter, which can be used to deduce information on the type and nature of chemical bonds in crystals.

We calculated the distribution of the potential in the (110) and (100) planes of GaP from the x-ray diffraction f curves found experimentally in [1]. The distribution of the potential can be represented in the form of a triple Fourier series [2, 3]

$$\varphi(x, y, z) = \frac{1}{V} \sum_{hkl} G_{hkl} e^{-2\pi i (hx + ky + lz)} \ ,\tag{1}$$

where the coefficients G_{hkl} are the structure amplitudes in electron diffraction. Using the relationship between the structure amplitudes in x-ray and electron diffraction, $G_{hkl} = F_{hkl} / \pi H^2$, which follows from Poisson's equation [3], we can write Eq. (1) in the form

$$\varphi(x, y, z) = \frac{1}{V\pi} \sum_{hkl} \frac{F_{hkl}}{H^2} \cdot e^{-2\pi i (hx + ky + lz)}\tag{2}$$

This series converges much faster than the Fourier series employed in calculations of the electron density distributions [3–5]. However, in view of the limited number of x-ray reflections, the present series cannot be computed by the usual summation method without committing an error in the termination procedure

We calculated the distribution of the potential using the experimentally determined f curves of the Ga and P ions, which were approximated by expressions of the type

$$f(\mu) = \sum_{i=1}^{n} \frac{A_i}{(1 + \alpha_i^2 \cdot \mu^2)^2} \ ,\tag{3}$$

where A_i and α_i are the approximation parameters; $\mu = 4\pi (\sin \vartheta)/\lambda$.

* "Semiconductors," pp. 135–142 (see page 3). This article was not presented at the International Symposium on Chemical Bonds in Semiconducting Crystals, Minsk, 1967.

The distribution of the potential of an atom or an ion expressed in spherical coordinates in terms of the atomic electron-scattering functions is [6]

$$\varphi'(r) = \frac{1}{K \cdot 2\pi^2 r} \int_0^\infty \mu f_{el}(\mu) \sin(\mu r) \cdot d\mu, \tag{4}$$

where $K = 2\pi me/h^2$. Using the well-known Mott formula [9]

$$f_{el} = \frac{8\pi^2 me^2}{h^2} \left(\frac{Z - f_x(\mu)}{\mu^2} \right), \tag{5}$$

which expresses the dependence of the atomic electron-scattering function f_{el} on the atomic x-ray scattering function f_x, we can use Eq. (4) to calculate the distribution of the potential of a given atom or ion. Substituting Eq. (5) in Eq. (4), we obtain

$$\varphi'(r) = \frac{eZ}{r} - \frac{2e}{\pi r} \int_0^\infty \frac{f_x(\mu) \sin(\mu r)}{\mu} \, d\mu, \tag{6}$$

where Z is the atomic number of the element in question.

The distributions of the potential for the Ga and P ions can be deduced from Eq. (3):

$$\varphi'(r) = \frac{eZ}{r} - \frac{2e}{\pi r} \sum_i^n \int_0^\infty \frac{A_i \sin(\mu r)}{(1 + \alpha_i^2 \mu^2)^2 \mu} \, d\mu. \tag{7}$$

Carrying out the integration and making the necessary calculations, we obtain

$$\varphi'(r) = \frac{e}{r} \left\{ Z - \sum_i^n A_i + \sum A_i e^{-r/\alpha_i} \left(1 + \frac{r}{2\alpha_i} \right) \right\}. \tag{8}$$

In order to determine the potential at any point in the lattice, we must sum the values of the potentials of all the atoms or ions which make a contribution at this point.

In the case of gallium phosphide, we obtain

$$\varphi(r) = \sum_j^{2N} \varphi'(r) = e \left\{ \left[\sum_j^N \frac{1}{r_{a_j}} \left(Z_A - \sum_i^n A_i \right) + \right. \right.$$

$$+ \sum_j^N \frac{1}{r_{a_j}} \sum_i^N A_i e^{-r_{a_j}/\alpha_i} \left(1 + \frac{r_{a_j}}{2\alpha_i} \right) \right] +$$

$$\left. + \left[\sum_j^N \frac{1}{r_{b_j}} \left(Z_B - \sum_i^n B_i \right) + \sum_j^N \frac{1}{r_{b_j}} \sum_i^n B_i e^{-r_{b_j}/\beta_i} \left(1 + \frac{r_{b_j}}{2\beta_i} \right) \right] \right\}, \tag{9}$$

where Z_A and Z_B are the atomic numbers of Ga and P, respectively; r_{a_j} and r_{b_j} are the distances of the Ga and P ions from the point at which the potential is being calculated; A_i, α_i, B_i, and β_i are the approximation parameters of the f curves of Ga and P, respectively; N is the number of atoms of a given kind considered in the calculation.

The experimental f curve of Ga was approximated by the sum of two curves of the type given by Eq. (3), whereas the f curve of P was approximated by the sum of three curves of the same type. The approximation parameters for Ga were $A_1 = 25.77$, $A_2 = 4.53$, $\alpha_1 = 0.086$, and $\alpha_2 = 0.374$, whereas for P they were $B_1 = 8.81$, $B_2 = 3.09$, $B_3 = 3.80$, $\beta_1 = 0.0614$, $\beta_2 = 0.306$, and $\beta_3 = 0.412$.

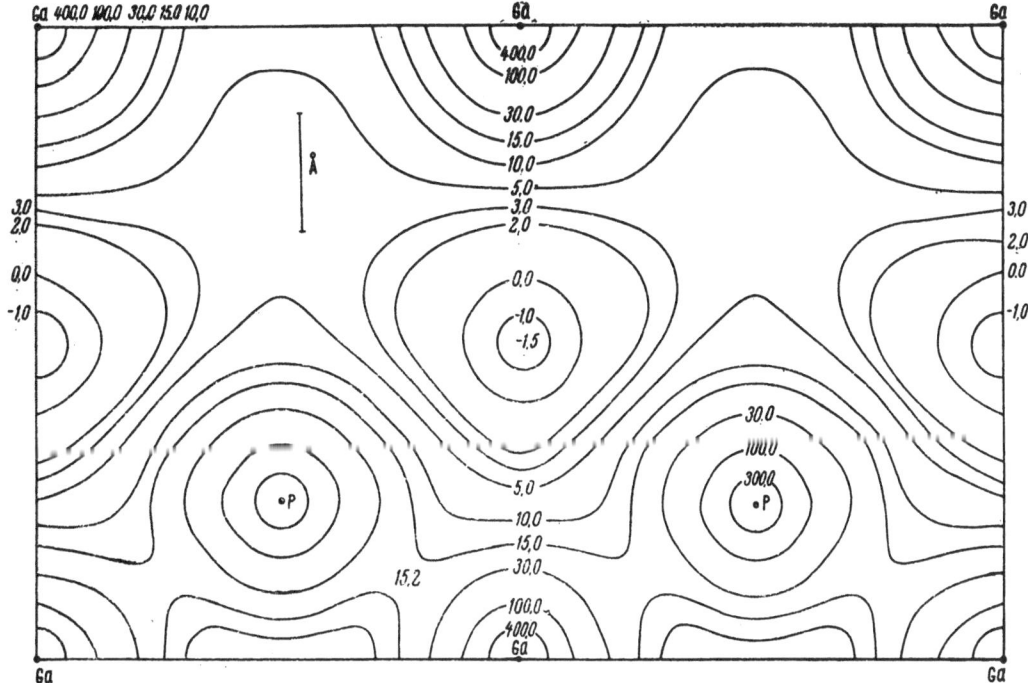

Fig. 1. Distribution of the potential in the (110) plane of GaP.

The potential in the (110) and (100) planes of the GaP lattice was calculated on a Minsk–22 computer. The program was tested by comparing the results of the manual and the machine calculations of the distribution of the potential at some points in the lattice, making allowance for the presence of atoms in only one unit cell. The two sets of results agreed to within a few tenths of a percent.

The convergence of the series represented by Eq. (9) was tested on a computer by calculating the value of the potential at some points along the [111] direction, including different numbers of coordination spheres. The calculations were carried out in such a way that the points at which the potential was computed were located at the centers of cubes of edges a, $3a$, $5a$, and $7a$, i.e., they were carried out for 1, 27, 125, and 343 unit cells. It was found that saturation occurred when the number of cells reached 8 and a further increase in the number of coordination spheres had practically no influence on the value of the potential. Therefore, the final calculations of the potential in the (110) and (100) planes of GaP were carried out making allowance for the presence of all the atoms in 27 unit cells at each point on the plane.

The edge of a unit cell was divided into 40 parts. The computer program was prepared in such a way that all the points at which the potential was calculated were located at the centers of cubes of edge $2a$, i.e., when the coordinates of the point were changed by $a/40$, the range of the calculations was shifted by the same amount. Suitable coefficients were introduced to ensure that the number of A atoms included in this was equal to the number of B atoms.

Figures 1 and 2 show the distributions of the potential in volts in the (110) and (100) planes. Figure 3 shows the distributions of the potential along the [111] and [001] directions.

It is evident from Figs. 1 and 2 that there were some negative potential regions in the (110) plane of GaP. The minimum value of the potential in the region of $\frac{1}{2}$, $\frac{1}{2}$, $\frac{1}{2}$ was −1.50 V. The negative value of the potential in this region was evidently due to the influence of the nearest negative P ions lying in the transverse plane. The potential between the nearest Ga and P atoms along the [111] directions rose to 15.21 V. The curve describing the distribution of the

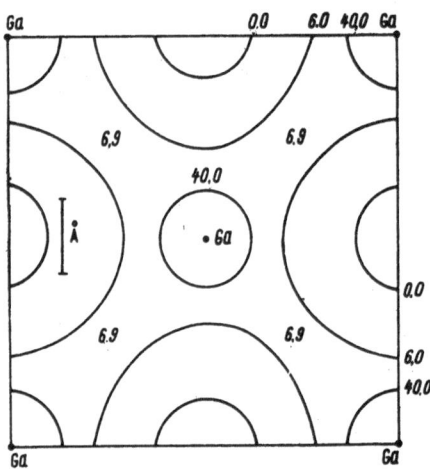

Fig. 2. Distribution of the potential in
the (100) plane of GaP.

potential along the [111] axis was asymmetric (Fig. 3a) with a tendency for the minimum to shift toward the phosphorus ion on either side, i.e., the steepness of the curve on the Ga ion side was somewhat less than on the P ion side. The minimum value of the potential between the nearest Ga ions along the [110] direction (Fig. 2) was 6.9 V.

We determined the value of the average internal potential in the GaP lattice. Knowing the potentials of the individual ions and their number in a unit cell, we could easily find the value of the average internal potential in the lattice.

If φ'_A (r) and φ'_B (r) in Eq. (8) represent the potentials of single A and B ions, the value of the average internal potential can be calculated from the formula

$$V_0 = \frac{n \int_V \varphi'_A (r)\, dV + m \int_V \varphi'_B (r)\, dV}{V}, \tag{10}$$

where n and m are, respectively, the numbers of A and B atoms in a unit cell, and V is the volume of one unit cell. Substituting Eq. (8) into Eq. (10) and integrating, we obtain

$$V_0 = \frac{8\pi e\, (n \sum A_i \cdot \alpha_i^2 + m \sum B_i \cdot \beta_i^2)}{V}. \tag{11}$$

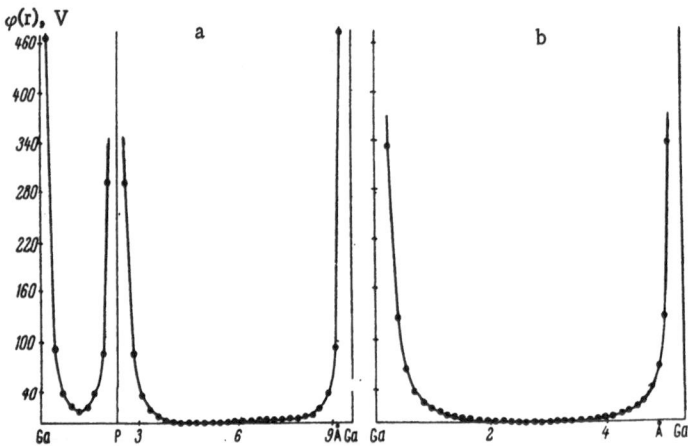

Fig. 3. Distribution of the potential along the [111] (a)
and [001] (b) directions in GaP.

Bearing in mind that, in our case, n = m = 4 and expressing the potential in volts, we find that

$$V_0 = \frac{1447.6 \left(\sum A_i \cdot \alpha_i^2 + \sum B_i \cdot \beta_i^2 \right)}{V}. \tag{12}$$

The average internal potential in gallium phosphide, calculated from Eq. (12), is 16.02 V.

It is known [7] that the average internal potential can be calculated also from the formula

$$V_0 = \frac{16\pi mc^2}{Ne} \cdot \frac{\chi_d}{V}, \tag{13}$$

where χ_d is the diamagnetic susceptibility.

Substituting into Eq. (13) the diamagnetic susceptibility of GaP, which is $-55 \cdot 10^{-6}$ [8], we obtain $V_0 = 14.5$ V.

The diamagnetic susceptibility χ_d, expressed in terms of the electron density distribution $\rho(r)$, is [7]

$$\chi_d = -\frac{4\pi e^2 N}{6mc^2} \int_0^\infty \overline{\rho(r)}\, r^4\, dr. \tag{14}$$

If the f curve is approximated by an expression of the type given by Eq. (3), we find that the electron density distribution for one ion is

$$\rho(r) = \sum_i \frac{A_i}{8\pi\alpha_i^3}\, e^{-r/\alpha_i}. \tag{15}$$

Substituting Eq. (15) in Eq. (14) and integrating over a lattice containing two types of ion, we obtain

$$\chi_d = -\frac{2Ne^2}{mc^2} \left(\sum_i A_i \cdot \alpha_i^2 + \sum B_i \cdot \beta_i^2 \right) = -3.38 \cdot 10^{-5} \left(\sum_i A_i \cdot \alpha_i^2 + \sum B_i \cdot \beta_i^2 \right). \tag{16}$$

The diamagnetic susceptibility χ_d of GaP, calculated from Eq. (16), is $-60.5 \cdot 10^{-6}$.

The authors are grateful to T. D. Sokolovskii for his help in the preparation of the computer program.

Literature Cited

1. N. N. Sirota and A. U. Sheleg, Vestsi Akad. Navuk Belarus. SSR, Ser. Fiz.-Mat. Navuk, No. 5, A103 (1968).
2. W. E. Lashkarev (Laschkarew), Phys. Z. Sowjetunion, 8:227 (1935).
3. S. G. Konobeevskii, Dokl. Akad. Nauk SSSR, 59:33 (1948).
4. N. N. Sirota, N. M. Olekhnovich, and A. U. Sheleg, Dokl. Akad. Nauk SSSR, 132:160 (1960).
5. R. W. James, The Optical Principles of the Diffraction of X Rays, Bell, London (1950).
6. B. K. Vainshtein, Structure Analysis by Electron Diffraction, Pergamon Press, Oxford (1964).
7. N. N. Sirota, E. M. Gololobov, N. M. Olekhnovich (Olechnovič), and A. U. Sheleg (Šeleg), Krist. Tech., 1:545 (1966).
8. Ya. G. Dorfman, Diamagnetism and Chemical Bond, American Elsevier, New York (1965).
9. N. F. Mott, Proc. Roy. Soc., London, 127:658 (1930).

PART II

THERMODYNAMIC AND THERMOCHEMICAL INVESTIGATIONS

PHASE EQUILIBRIA AND TRANSITIONS
AT LOW TEMPERATURES*

V. K. Semenchenko

The theorem of the ultimately stable states is formulated and its consequences are briefly enumerated. This theorem is used to derive the properties of ultimately stable phases and the special features of an equilibrium between ultimately stable and normal phases. Possible phase transitions between them, including the critical transition, are considered. The special nature of the behavior of the transport coefficients in ultimately stable phases is considered. The general conclusions are tested against the properties of superconductors and of a supercaloric phase characterized by zero entropy and temperature. The experimental data for superconductors are in agreement with the general formulas. The application of these formulas to the supercaloric phase shows that the Nernst postulate is a special case of the theorem of ultimately stable states. Since the state of zero entropy and temperature is a new phase, it can be reached only after a phase transition. This transition should make it possible to reach the absolute zero if the thermal expansion coefficient is positive. However, the low-temperature values of the thermal expansion coefficient of many substances are negative. Therefore, we may assume that substances can be divided into two groups: those with a positive value of $(\partial V/\partial T)_p$, which may go over to the super-caloric state, and those with a negative value of $(\partial V/\partial T)_p$, which cannot reach this state. It is also possible that a reversal of the sign of $(\partial V/\partial T)_p$ near $T = 0°K$ is a general law and, therefore, the super-caloric state cannot be reached.

Phase equilibria and transitions at low temperatures are best considered on the basis of the theorem of ultimately stable states [1] because at these temperatures matter is either in the ultimately stable state or is close to it. Let us recall briefly the theorem and its consequences. A system is in thermodynamic equilibrium if the Jacobian D_y (made up of the second derivatives of the internal energy U) and all its diagonal minors are positive, i.e.,

$$D_y = \frac{D(X_1 \ldots X_n)}{D(x_1 \ldots x_n)} > 0; \quad \frac{D(X_1 \ldots X_{n-1})}{D(x_1 \ldots x_{n-1})} > 0; \quad \ldots \left(\frac{\partial X_i}{\partial x_i} \right)_{x_j} > 0 \qquad (j \neq i), \tag{1}$$

where

$$X_i = \left(\frac{\partial U}{\partial x_i} \right)_{x_j}; \quad \left(\frac{\partial X_i}{\partial x_i} \right)_{x_j} = \left(\frac{\partial^2 U}{\partial x_i^2} \right)_{x_j}.$$

However, D_y, like any other positive quantity, must lie within the limits

$$0 < D < +\infty. \tag{2}$$

Without violating the condition that the minors should be finite and positive, we can reach the upper limit of $+\infty$ by making one of the diagonal terms of the Jacobian the coefficient of stability

---

* "Crystals," pp. 213-219 (see page 3).

$(\partial X_k / \partial x_k)_{x_i} \to \infty$. Then, expanding D_y in elements of the first row containing $(\partial X_k / \partial x_k)_{x_i} \to \infty$, we can satisfy the condition (2). The limit $D_y = 0$ (a spinode) has already been considered by Gibbs [2] and, therefore, we shall ignore it.

In order to deal with the consequences of our assumption that we can reach the upper limit of stability, we shall consider a Jacobian D_y' which is the reciprocal of D_y:

$$D_y' = \frac{D(x_1 \dots x_n)}{D(X_1 \dots X_n)} = \frac{1}{\dfrac{D(X_1 \dots X_n)}{D(x_1 \dots x_n)}} = \begin{vmatrix} \left(\dfrac{\partial x_1}{\partial X_1}\right)_{X_i} \cdots \left(\dfrac{\partial x_1}{\partial X_n}\right)_{X_i} \\ \cdot \; \cdot \; \cdot \; \cdot \; \cdot \; \cdot \; \cdot \; \cdot \; \cdot \; \cdot \\ \left(\dfrac{\partial x_n}{\partial X_1}\right)_{X_i} \cdots \left(\dfrac{\partial x_n}{\partial X_n}\right)_{X_i} \end{vmatrix}_{D_y \to \infty} = 0. \qquad (3)$$

This condition can be satisfied simply by making the elements of one of the rows of the Jacobian vanish. Since all the quantities $(\partial X_i / \partial x_i)_{x_j}$ are the second derivatives of the total differential of U, the Jacobian of Eq. (3) is symmetrical and, therefore, not only the elements of the first row but also those of the first column vanish and we have

$$\left(\frac{\partial x_1}{\partial X_i}\right)_{X_j} = 0, \quad \left(\frac{\partial x_i}{\partial X_1}\right)_{X_j} = 0. \qquad (4)$$

Consequently, in the system which satisfies the minimum conditions necessary for the realization of the ultimately stable state $D_y \to \infty$, all the derivatives vanish which contain either the force X_k or the coordinates x_k, occurring in the stability coefficient $(\partial X_k / \partial x_k)_{x_i} \to \infty$:

$$\left(\frac{\partial x_1}{\partial X_1}\right)_{X_i}, \quad \left(\frac{\partial x_2}{\partial X_1}\right)_{X_i}, \quad \dots, \quad \left(\frac{\partial x_n}{\partial X_1}\right)_{X_i} = 0;$$

$$\left(\frac{\partial x_1}{\partial X_2}\right)_{X_i}, \quad \left(\frac{\partial x_1}{\partial X_3}\right)_{X_i}, \quad \dots, \quad \left(\frac{\partial x_1}{\partial X_n}\right)_{X_i} = 0.$$

All the other derivatives can have any finite value.

We shall use the term characteristic parameters for the quantities occurring in $(\partial X_k / \partial x_k)_{\bar{x}_i} \to \infty$. The total differential of the characteristic coordinate follows from Eq. (4):

$$dx_k = \sum_i \left(\frac{\partial x_k}{\partial X_i}\right)_{X_j} dX_i = 0, \quad x_k = \text{const.} \qquad (5)$$

Consequently, in the ultimately stable state, the characteristic coordinate x_k has a constant value, i.e., it loses its ability to influence any of the generalized forces, and the characteristic force is no longer capable of altering any of the thermodynamic coordinates. The system as a whole loses the properties which are described by the characteristic forces and coordinates. Therefore, these properties may be assumed to vanish because zero represents, basically, the absence of any value.

Let us consider two examples. First, we assume that $X_k = H$, $x_k = B$, and $X_1 = T$, $X_2 = -p$, $x_1 = s$, and $x_2 = V$. Then, the expressions

$$\left(\frac{\partial H}{\partial B}\right)_{s,V} = \frac{1}{\mu} \to \infty, \quad \left(\frac{\partial B}{\partial H}\right)_{T,p} = \mu = 0,$$

$$\left(\frac{\partial B}{\partial T}\right)_{H,p} = \left(\frac{\partial s}{\partial H}\right)_{T,p} = 0,$$

$$-\left(\frac{\partial B}{\partial p}\right)_{T,H} = \left(\frac{\partial V}{\partial H}\right)_{T,p} = 0 \qquad (6)$$

give all the equilibrium thermodynamic properties of superconductors or, more correctly, of superdiamagnetic substances: according to these expressions, the quantities that vanish are the magnetostriction $(\partial V/\partial H)_{T,p}$, the magnetocaloric effect $T(\partial s/\partial H)_{T,p}$, the magnetic induction B, etc.

In the second example, we shall assume that $X_k = T$ and $x_k = S$. We then obtain the following expressions

$$\left(\frac{\partial T}{\partial S}\right)_{p,V} \to \infty, \quad \left(\frac{\partial S}{\partial T}\right)_{p,H} = \frac{C_{p,H}}{T} = 0,$$

$$\left(\frac{\partial S}{\partial H}\right)_{T,p} = \left(\frac{\partial B}{\partial T}\right)_{p,H} = 0,$$

$$-\left(\frac{\partial S}{\partial p}\right)_{T,H} = \left(\frac{\partial V}{\partial T}\right)_{p,H} = 0, \quad S = 0,$$

i.e., in its ultimately stable state, the system possesses all the properties which follow from the Nernst postulate in the state of zero temperature and entropy (T = 0, S = 0). However, if the specific heat C_{x_i} is a linear function of the temperature, the ultimately stable state cannot be reached because C/T = const, i.e., the theorem of ultimately stable states imposes restrictions which are more stringent than the limitations of the Nernst postulate because not only must $(\partial S/\partial T)_x$ vanish but also

$$\lim_{S \to 0}\left(\frac{\partial^2 S}{\partial T^2}\right) = 0.$$

All the derivatives which contain the characteristic constants are inversely proportional to the fluctuations and, therefore, they can tend to infinity only when the amplitude of the fluctuations is zero. Consequently, an ultimately stable system is in a state of complete determinacy. This state can only be the lowest quantum state. In the presence of degeneracy, which may be lifted by external fields, the determinacy is violated and some field is capable of taking the system out of its ultimately stable state.

Possibility of Reaching Absolute Zero

Superdiamagnetic substances and systems remaining at absolute zero are the two special cases of ultimately stable states ("superstates") and, therefore, they should obey the same laws. The conditions for a phase equilibrium between a superphase and a normal phase are:

$$Z' = Z'', \quad U_C - \sum_i X_i^C x_i^C = U - X_k x_k - \sum_{i \neq k} X_i x_i. \tag{7}$$

The conditions of Eq. (7) show that we are dealing with a phase equilibrium of a new type which cannot be realized under normal conditions: a characteristic force X_k has different values in the two phases, which are $X_k = 0$ in a superphase and $X_k \neq 0$ in a normal phase. This is known to apply to superconductors for which the difference between the equilibrium values of the magnetic field in the case of "hard" materials may reach 10^5 Oe. It is strange that this violation of the laws of thermodynamic equilibrium has not yet attracted special attention. Since superconductors satisfy the conditions of Eq. (7) it follows that a superdiamagnetic state may be reached when an external force X_k does not tend to zero but has a finite (and even very large) value. In this case, the temperature field can transform the system to a superstate. Hence, we may conclude that there is no need to reduce the temperature to T = 0 in order to reach the supercaloric state S = 0, T = 0. External (for example, mechanical) forces should be sufficient to transform the system to an ultimately stable state. In this process, the condition $(\partial s/\partial p)_T < 0$ should be obeyed. However, we know that $(\partial s/\partial p)_T = -(\partial V/\partial T)_p$; consequently, if the temperature coefficient $(\partial V/\partial T)_p$ is positive, an increase in the pressure should reduce the entropy of the system and this should make it possible to reach the supercaloric state. The usual objections against the possibility of reaching the state $S = 0$, $T = 0$ lose their force because we are dealing with a completely new phase transition predicted by the theorem of ultimately stable states. However, it is known that many substances (ice, He, Si, Ge, α-Sn [3−6]) exhibit a reversal of the sign of $(\partial V/\partial T)_p$ near $T = 0$ so that this quantity approaches asymptotically the limiting value $(\partial V/\partial T)_p = 0$ from the negative side. Therefore, an increase in the pressure will increase the entropy of the system S, i.e., we would have to use expansion to reduce the entropy. Alternatively, if the negative sign of this derivative is a general property, we may regard this property as a barrier preventing the realization of the supercaloric state. If $(\partial V/\partial T)_p$ is positive, we can calculate the temperature and the pressure corresponding to equilibrium between the supercaloric and normal phases:

$$U_c + pV_c = U - TS + pV, \quad T = \frac{U - U_c + p(V - V_c)}{S}. \tag{8}$$

A supercaloric phase has mechanical and electromagnetic but no thermal properties. A supermechanical phase, i.e., a perfectly hard body, with $(\partial V/\partial p)_X = 0$, is impossible to achieve because the velocity of propagation of any mechanical disturbance in such a body would be infinite.

Limited Phase Transitions of the Second Kind

Phase transitions of the second kind are possible only in systems in which one of the phases is in a superstate [7]. We shall show that, in this case, the Ehrenfest postulate is satisfied by the second derivatives of the Z potential which contain either one or both characteristic parameters x_k and X_k. Let us consider the final point of the equilibrium between a superphase and a normal phase. Since the conditions for the existence of a superphase $(\partial X_k/\partial x_k)_{x_j} \rightarrow \infty$, $x_k = 0$ cannot be satisfied by a normal phase it follows that X_k at the final point must be zero for both phases. The other coordinates and forces (X_i, x_i) should be equal in both phases, i.e., the conditions of existence of a critical point should be satisfied, with the exception of the characteristic coordinates of the two phases. Differentiation of the condition of equilibrium $Z_C = Z'$, which now becomes an identity, gives:

$$d(Z_c - Z') = x' + \sum_{i \neq k} (x_c - x')\, dx' = -x',$$

$$\tag{9}$$

$$d^2(Z_c - Z') = \sum_i \left(\frac{\partial^2 Z_c}{\partial x_h \partial x_i} - \frac{\partial^2 Z'}{\partial x_k' \partial x_i'} \right) dx_h dx_i + \sum_i \left(\frac{\partial^2 Z_c}{\partial x_i \partial x_j} - \frac{\partial^2 Z'}{\partial x_i \partial x_j} \right) dx_i dx_j.$$

The Ehrenfest postulate is now satisfied only by $\partial^2 Z/\partial x_i \partial x_k$, which contains either X_k or x_k, or both. The other second derivatives are equal in both phases, as at a normal critical point. The minor $D(X_1 \dots X_{n-1})/D(x_1 \dots x_{n-1})$ behaves as in normal phase transitions. Consequently, the ultimate phase transition in systems containing a superphase represents a superposition of a normal critical transition and a special transition characterized by the discontinuities of all the second derivatives $\partial^2 Z/\partial X_i \partial X_k$ which contain the characteristic parameters (these derivatives change discontinuously from zero in a superphase to values typical of a normal phase). Since, in contrast to x_i, we cannot reach the value $x_k = 0$ without transition to a superstate, the ultimate transition to a supercaloric state is impossible and the transition to the superdiamagnetic (superconducting) state is the only known phase transition which satisfies partly the Ehrenfest postulate. Therefore, this and other similar transitions can be called limited phase transitions of the second kind.

It is demonstrated in [8] that the transport coefficients (thermal diffusivity, diffusion coefficient, fluidity, etc.) considered in the Fourier approximation are proportional to the stability coefficients. This makes it possible to determine whether we are dealing with a critical transition or a limited phase transition of the second kind and, in the latter case, which of the parameters are characteristic. In critical transitions, the transport coefficients decrease strongly, whereas in limited transitions of the second kind they tend to infinite values. This criterion shows that phase transitions of the second kind which occur in binary alloys, polymers, ferromagnets, ferroelectrics, liquid crystals, etc., are essentially transcritical transitions, which are sometimes close to the critical conditions because the values of the transport coefficients decrease strongly at the transition point. The occurrence of superfluidity in He II demonstrates that, even in the absence of a coordinate or a derivative which tends to zero, this substance is a superphase in the kinetic sense.

Literature Cited

1. V. K. Semenchenko, Izv. Akad. Nauk SSSR, Otd. Khim. Nauk, No. 2, p. 368 (1959); No. 11, p. 2048 (1959); Zh. Fiz. Khim., 33:1440 (1959).
2. J. W. Gibbs, Collected Works, Vol. 1, Longmans, New York (1931).
3. S. Valentiner and J. Wollot, Ann. Phys. (Leipzig), 46:837 (1945).
4. H. D. Erfling, Ann. Phys. (Leipzig), 41:467 (1942).
5. H. Andenstedt, Ann. Phys. (Leipzig), 26:69 (1936).
6. S. I. Novikova, Fiz. Tverd. Tela, 2:1617 (1960); 2:43 (1960); 2:2341 (1960); 3:178 (1961); S. I. Novikova and N. Kh. Abrikosov, 5:2138 (1963).
7. V. K. Semenchenko, Zh. Fiz. Khim., 34:1649 (1960).
8. V. K. Semenchenko, Zh. Fiz. Khim., 34:1384 (1960).

THE AVERAGE HEAT OF ATOMIZATION AND
THE PROPERTIES OF SEMICONDUCTORS*

V. Sadagopan and H. C. Gatos

*Departments of Metallurgy and Electrical Engineering
and Center for Materials Science and Engineering
Massachusetts Institute of Technology
Cambridge, Massachusetts*

In correlating the physical properties of known semiconductors and predicting the properties of new ones a number of parameters have been employed with varying degrees of success. Typical parameters have been the electronegativity difference, the heats of formation, the average principal quantum number and the atomic or ionic radii. All such parameters are related to the binding energy of semiconductors and consequently to properties such as their energy gap. In the present study the cohesive energy was considered as the most basic parameter for analyzing the characteristics of semiconductors. In elemental semiconductors the cohesive energy is given by the energy of atomization and is known in all instances. In contrast, the cohesive energy of compound semiconductors is not generally known. A simple method of calculating the heats of atomization of compounds from the heats of atomization of the elements was developed. The resulting "average heats of atomization" do not contain the contribution of the heats of formation but they can serve essentially as the cohesive energies particularly in predominantly covalent semiconductors, where the heats of formation represent only a small fraction of the cohesive energy. The "average heats of atomization" were found to be very successful in correlating the properties of a number of groups of semiconductor compounds including the vitreous semiconductors and the limits of stability of various structures among semiconductor compounds.

Introduction

The striking advances in solid state science and technology since the discovery of the transistor in 1948 have hinged on the development of new materials of the appropriate composition of high crystalline perfection and of controlled chemical purity. As in the past, today also the slow step in the progress of solid state is the availability of the appropriate materials with the desired purity or with the desired level of the proper impurities.

The energy band approximation constitutes a rigorous means for the understanding of the electronic behavior of semiconductors. However, in view of the complexity of the many electron systems the energy band approximation has been meaningfully or fully carried out for only a few simple systems. Furthermore, the band approximation does not lead to relationships concerning the dependence of electrical properties of materials on the chemical composition. It is

---

* "Crystals," pp. 220-231 (see page 3).

this type of relationship that is essential for the design, development and preparation of needed new electronic materials.

In contrast to the band approximation, the chemical or bond approximation is concerned with the interatomic forces primarily between neighboring atoms, and neglects the longer range interactions. Obviously, it does not lend itself to quantitative refinement. However, in relating the strength of the bonds to the chemical nature of the participating atoms, the chemical bond approach [1] leads to relatively simple and convenient correlations of various material properties and consequently to predictions concerning new materials.

A number of criteria have been used, with varying degrees of success, in assessing the bond strength in semiconductors and, thus, correlating the electrical properties to chemical composition and designing new materials of the desired nature. Among the criteria employed are the electronegativity difference, the melting point, the heat of formation, the average principal quantum number, and others [2].

In the present report the "average heats of atomization" will be employed as a measure of the bond strengths and, thus, as a criterion for correlating the electrical properties of semiconductors. The "average heats of atomization" will be further discussed as a criterion for correlating the limits of phase stability in some semiconductor systems.

Average Heats of Atomization

The heat (enthalpy) of atomization of the elements represents the cohesive energy and consequently the associated bond strength. In the case of compounds the heats of atomization represent again the cohesive energy and thus reflect the relative bond strength among isostructural materials. The heats of atomization have been used by Pauling for the computation of bond strengths [3]. They have also been employed in conjunction with the energy gaps in some semiconductors with the tetrahedral coordination [4]. The heats of atomization of the elements have been employed further in correlating their crystalline structures and the structure of some intermetallic phases [5].

For numerous semiconductor compounds, experimental values of the heats of atomization (H_S) are not available. We have proposed [6] that an average heat of atomization ($\overline{H}_S$) can be employed as a parameter of relative bond strength and consequently as the basis for correlating the electrical properties of semiconductors and their chemical composition. For a compound $A_x B_y C_z$, this parameter $\overline{H}_S$ in kcal/g-atom is defined as

$$\overline{H}_s = \frac{xH_S^A + yH_S^B + zH_S^C}{x+y+z},$$ (1)

where H_S^A, H_S^B, and H_S^C are the heats of atomization of the elements A, B, and C, respectively.[1] It is apparent that the parameter $\overline{H}_S$ does not contain the contribution of the heat of formation (H_F) to the cohesive energy of a compound. In fact it is expected that

$$H_S = H_F + \overline{H}_S.$$ (2)

In Eq. (2) H_S represents the total bond energy, H_F represents the ionic contribution, and $\overline{H}_S$

---

[1] The heat of atomization of element A is taken to represent the energy in kcal/g-atom associated with the reaction $A_{(s)} \rightleftharpoons A_{(g)}$ at 25°C, where (s) stands for solid and (g) for monatomic gas. For elements which are gases or liquid at room temperature, the heats of atomization correspond to the temperature of the revelant melting point.

TABLE 1. Comparison of Heats of Formation (H_F, in kcal/g-atom), Average Heats of Atomization ($\bar{H}_S$, in kcal/g-atom), Experimental Heats of Atomization (H_S, in kcal/g-atom), Energy Gaps (E_g, at 300°K, in eV), Electronegativity Differences (Δx), Molecular Weight (M. W.), Melting Point (M. P., in °K) and Ratios of Cation to Anion Radius (r_c/r_a)

Compound	H_F	$\bar{H}_S$	$\bar{H}_S+H_F$	H_S [9]	E_g [10]	Δx	M.W.	M.P. [10]	r_c/r_a
AlP		78.4			3 0	0.6	57.96	1770	0.87
GaP		74.4			2.24	0.5	100.7	1623	0.88
AlAs		73.3			2.16	0.5	101.89	1870	0.94
AlSb	1.5 [7]	69.8	71.3		1.6	0.4	148.74	1327	1.08
GaAs		69.0			1.35	0.4	144.63	1510	0.94
InP		68.9			1 27	0.4	145.8	1327	0.76
GaSb	2.49 [7]	65.5	67.99		0.81	0.3	191.48	985	1.08
InAs	3.7 [8]	63.5	67.2	71.5	0.36	0.3	189.73	1216	0.82
InSb	1.7 [8]	60.0	61.7	57.0	0.18	0.2	236.56	798	0.94

TABLE 2. Comparison of Heats of Formation (H_F, in kcal/g-atom), Average Heats of Atomization ($\bar{H}_S$, in kcal/g-atom), Experimental Heats of Atomization (H_S, in kcal/g-atom), Energy Gaps (E_g, at 300°K, in eV), Melting Points (M. P., in °C), Microhardness (mh, in kg/mm²), Debye Temperatures (θ, in °K at 80°K), Refractive Indices (n), and Average Principal Quantum Numbers ($\bar{Q}$)

Compound	H_F [11]	H_S	H_S+H_F	H_S [11]	E_g [10]	M. P.	mh [7]	θ [10]	n [7]	$\bar{Q}$
β-ZnS	24.3	48.6	72.9	73.0	3.54	1923			2.37	3.5
HgS	6.9	40.7	47.6	47.0		1750 (120 atm)			2.85	4.5
ZnSe	19.5	40.3	59.8	63.0	2.58	1788±20	135±5	400	2.89	4.0
ZnTe	13.0	38.6	51.6	52.0	2.25	1512	90±5	250	3 56	4.5
β-CdTe	12.0	36.4	48.4	48.0	1.44	1371±2	60±5	200	2.5	5.0
HgSe	7.0	32.4	39.4	42.0	0.3	1038±5		212		5.0
HgTe	6.0	30.7	36.7	37.0	0.15	928±5				5.5

TABLE 3. Comparison of Average Heats of Atomization ($\bar{H}_S$, in kcal/g-atom) and Energy Gaps (E_g, in eV) of Several Classes of Isostructural Materials

Compound	Structure	$\bar{H}_S$	E_g	Compound	Structure	$\bar{H}_S$	E_g
Mg₂Pb	C1	39.3	0.1 [13]	FeAs₂	C18	79.2	0.25 [15]
Mg₂Sn	C1	44.4	0.33 [10]	FeP₂	C18	86.4	0.4 [15]
Mg₂Ge	C1	53.7	0.74 [10]	RuSb₂	C18	93.0	>0.3 [15]
Mg₂Si	C1	59.7	0.77 [10]	RuPAs	C18	97.6	~0 8 [15]
Bi₂Te₃	C33	47.4	0.15 [14]	OsSb₂	C18	103.6	>0.3 [15]
Bi₂Se₃	C33	49.4	0.35 [14]	RuP₂	C18	104.9	~1.0 [15]
Sb₂Se₃	C33	54.2	1.2 [14]	OsAs₂	C18	108.3	0.9 [15]
FeSb₂	C18	74.5	0.05 [15]	OsP₂	C18	115.5	1.2 [15]

the covalent contribution. As is seen in Tables 1 and 2 the sum $\overline{H}_S + H_F$ is in excellent agreement with the experimentally determined values of the H_S. For convenience, the heats of atomization and the heats of formation are expressed throughout this report in kcal/g-atom rather than kcal/g-mole.

In most instances the heats of formation are not available. However, the $\overline{H}_S$ parameters still lead to very successful correlations since the H_F values usually follow the trend of the cohesive energies. In numerous compounds, of course, the heat of formation is relatively small to be of significance.

Comparison of $\overline{H}_S$ with Other Criteria

The computed average heats of atomization of the better-known III–V semiconductor compounds are listed in Table 1 together with several other parameters employed for correlating the properties of semiconductors and particularly the energy gaps. It is apparent that the $\overline{H}_S$ parameter is the most directly related to the trend of the energy gaps.

In many instances the heats of formation correlate well with the energy gap values since H_F essentially reflects the ionic part of bonding to which the energy gap is also distinctly related. However, there are many inconsistencies as, for instance, in the V_2-VI_3-type compounds. For Sb_2Te_3 the heat of formation (2.70 kcal/mole) is smaller than the heat of formation of Bi_2Te_3 (3.74 kcal/mole) [12] and yet the energy gap of Sb_2Te_3 (0.3 eV) is greater than that of Bi_2Te_3 (0.15 eV). In the present case as well as in all others where the heats of formation are not correspondingly consistent with the energy gaps, the average heats of atomization present no inconsistencies (thus, $\overline{H}_S$ is 52.4 for Sb_2Te_3 and 47.5 for Bi_2Te_3).

Although a most useful parameter, the electronegativity difference obviously lacks the sensitivity for correlating the energy gaps. The isostructural compounds Mg_2Si, Mg_2Ge, Mg_2Sn, and Mg_2Pb have the same electronegativity difference (0.6), whereas their energy gaps decrease from 0.77 eV to 0.1 eV (see also Table 3). Electronegativity is particularly inadequate for correlating the properties of ternary and of higher-order semiconductor compounds.

Similarly, the average principal quantum number $\overline{Q}$ is rather insensitive as shown in Table 2 where compounds with the same values of $\overline{Q}$ have distinctly different energy gaps.

As is shown in Table 1 the cation-to-anion ratio is limited in its applicability as a parameter for correlating the energy gaps of compound semiconductors. Other proposed criteria based on atomic and ionic radii or lattice parameters present similar limitations.

Crystalline Semiconductors

Physical Properties

The relationship between the average heats of atomization and the physical properties – notably the energy gap – of crystalline isostructural semiconductors becomes apparent in Tables 1–6. It is not clear as yet how one could correlate different classes of compounds with the same structure (for instance III–V and II–VI compounds with the zinc blende B3 structure) or different classes of compounds with different structures. Apparently there exists a high degree of specificity of the physical properties within the various classes of semiconductor compounds which cannot be resolved with our present understanding of the prevailing interatomic forces.

Phase Stability

The phase stability within certain classes of compounds can be related to the average heats of atomization. For example, a number of IV–VI and related compounds are listed in

V. SADAGOPAN AND H. C. GATOS

TABLE 4. Comparison of Average Heats of Atomization ($\overline{H}_S$, in kcal/g-atom) and Energy Gaps (E_g, in eV) of Some Chalcopyrites ($E1_1$ Structure) of General Formula II−IV−V_2

Compound	$\overline{H}_S$	E_g	Compound	$\overline{H}_S$	E_g
$ZnSiP_2$	74.7	—	$CdSnP_2$	64.6	1.5 [7]
$ZnGeP_2$	70.2	2.2 [7]	$CdGeAs_2$	63.7	0.53 [16]
$ZnSiAs_2$	69.3	2.1 [7]	$ZnSnAs_2$	60.3	—
$CdGeP_2$	69.1	1.8 [7]	$CdSnAs_2^*$	59.2	0.23 [16]
$ZnSnP_2^*$	65.7	—	$CdSnSb_2$	55.7	—
$ZnGeAs_2$	64.8	>0.6 [7]	—		—

*Obtained as two-phase substance.

TABLE 5. Comparison of Average Heats of Atomization ($\overline{H}_S$, in kcal/g-atom) and Energy Gap (E_g, as Determined by Photoelectric Measurements, in eV) of Compounds with General Formula II−III_2−VI_4 (Defect-Chalcopyrite Structure)

Compound	$\overline{H}_S$	E_g	Compound	$\overline{H}_S$	E_g
$ZnGa_2S_4$	61.9	—	$CdGa_2Se_4$	51.8	2.43 [17]
$CdGa_2S_4$	61.3	3.44 [17]	$HgGa_2Se_4$	50.1	1.95 [17]
$HgGa_2S_4$	59.6	2.84 [17]	$ZnIn_2Se_4$	49.3	1.82 [17]
$ZnIn_2S_4$	58.5	2.6 [17]	$CdIn_2Se_4$	48.6	1.72 [17]
$CdIn_2S_4$	58.1	2.3 [17]	$ZnIn_2Te_4$	47.3	0.86−1.15* [18,19]
$HgIn_2S_4$	56.5	2.0 [17]	$CdIn_2Te_4$	46.7	0.92−1.08* [18,19]
$ZnGa_2Se_4$	52.4	—	$HgIn_2Te_4$	45.0	0.86* [18.19]

*Optical energy gap values.

TABLE 6. Comparison of Average Heats of Atomization ($\overline{H}_S$, in kcal/g-atom), Microhardness (mh, in kg/mm^2), and Thermal Conductivity (κ, in cal·cm^{-1}·sec^{-1}·deg^{-1}) of Compounds with General Formula I_2−IV−VI_3 (B^3 Structure)

Compound	$\overline{H}_S$	mh [17]	k [17]	Compound	$\overline{H}_S$	mh [17]	k [17]
Cu_2SnS_3	72.0	283	—	$Cu_2GeSe_3·2Cu_2SnSe_3$	64.7		9.1·10^{-3}
$2Cu_2GeSe_3·Cu_2SnSe_3$	65.7		7.1·10^{-3}	Cu_2SnSe_3	63.7	256	—
$Cu_2GeSe_3·Cu_2SnSe_3$	65.2		8.5·10^{-3}	Cu_2SnTe_3	62.0	201	—

Table 7 in the order of increasing $\overline{H}_S$. It is seen that all compounds with $\overline{H}_S$ values less than 60.7 kcal/g-atom have the rocksalt (B1) structure. In contrast, compounds with $\overline{H}_S$ values greater than about 61 kcal/g-atom acquire various distorted forms of the B1 structure.

The rocksalt structure in the IV−VI-type compounds (like PbTe) and the associated p^3 bonding is possible if the filled s orbitals of both the Pb(s^{2}p^2) and the Te(s^{2}p^2) atoms act as nonbonding orbitals so that only six valence electrons are available for the bond formation. A similar type of bonding arrangement may prevail among the derivatives of the IV−VI compounds with the rocksalt structure. However, as the s and p orbitals of the cation become close in

TABLE 7. Comparison of Average Heats of Atomization ($\bar{H}_S$, kcal/g-atom), Energy Gap Values (E_g, in eV), and Crystal Structures of IV−VI and Related Compounds

Compound	Structure	$\bar{H}_S$	E_g	Compound	Crystal type or structure	$\bar{H}_S$	E_g
PbTe	B1	46.40	0.28 [10]	SnSe	B29	60.70	0.88 [22]
PbSe	B1	48.10	0.26 [10]	32% GeTe—68%SnTe	B1	60.88	—
KBiS$_2$	B1	50.75	—	1 : 1AgSbS$_2$: PbS**	monoclinic	61.0	—
NaBiS$_2$	B1	51.85	—	3 : 2AgSbS$_2$: PbS**	"	61 92	—
AgBiTe$_2$	B1	52.48	0.30 [20]	AgBiS$_2$***	orthorhombic	62.48	—
LiBiS$_2$	B1	54.98	—	PbSnS$_2$	B29	62.70	—
AgSbTe$_2$	B1	55.60	0.7* [20]	AgSbS$_2$	monoclinic	65.6	—
PbS	B1	56.40	0.38 [10]	AgAsS$_2$	"	67.35	—
AgSbSe$_2$	B1	57.30	0.6 [20]	GeTe	rhombohedral	68.00	1.0 [22,23]
AgAsTe$_2$	B1	57.35	0.8 [20]	SnS	B29	69.00	1.26 [22]
SnTe	B1	59.0	0 3 (4.2° K); 0.10 (000° I() [21]*	GeSe	orthorhombic	69.70	—
				GeTe	Lowe symmetry	77.0	—
AgAsSe$_2$	B1	59.10	0.9 [20]	GeS	B16	78.0	1.77 [22]
AgGeSbSe$_3$	B1	59 77	—	—		—	—

* Energy gap values from tunneling experiments. Recently it has been suggested that the valence and conduction bands of SnTe are inverted from those of PbTe, on the basis of the temperature dependence of their energy gaps [24].

** At higher temperatures, phase transformation to B1 structure takes place.

*** This value may be in error in view of the difficulties associated with preparing the material as single phase.

energy (as the atomic number of the cation decreases), then the sp^3 orbital hybridization in the cation is favored over the p^3 bonding. As a result, the rocksalt structure undergoes distortion. The normal tetrahedral sp^3 configuration cannot be realized since the total number of valence electrons (in a IV−VI molecule) is ten rather than the required eight. The two extra electrons have a pronounced destabilizing effect in a tetrahedral configuration [1].

As seen in Table 7 the $\bar{H}_S$ value of approximately 61 kcal/g-atom represents the transition point of the cohesive energies between the B1 structure and its distorted derivative structures. No specific significance need necessarily be assigned to this absolute value of $\bar{H}_S$.

Consistent with the above observations of the IV−VI and related compounds are the data listed in Table 8. At room temperature GeTe has a distorted B1 structure (rhombohedral) and an $\bar{H}_S$ value of 68.0 kcal/g-atom. At 725°C it transforms to the B1 structure. Since $\bar{H}_S$ decreases with increasing temperature it probably attains a value of less than 61 kcal/g-atom at 725°K. The rest of the phase transition temperature data of Table 8 confirm further the correlation between $\bar{H}_S$ and the limits of stability of the B1 structure.

Tables 9 and 10 represent two further cases, where the heats of atomization correlate with phase stability. Here, the electronegativity differences are relatively large and according-

TABLE 8. Comparison of Average Heats of Atomization ($\bar{H}_S$, in kcal/g-atom) and Transition Temperatures (T_{tr}, in °K) for the Rhombohedral-to-Cubic (B1) Transformation among Germanium and Tin Tellurides

Composition	$\bar{H}_S$	T_{tr}	Composition	$\bar{H}_S$	T_{tr}
GeTe	68.0	725 [25]	Ge$_{0.9}$Tl$_{0.05}$Bi$_{0.05}$Te	65.82	500 [25]
96GeTe — 4PbTe	67.14	665 [25]	17GeTe—83SnTe	61.53	295 [25]
95GeTe — 5Bi$_2$Te$_3$	66.97	575 [25]	10GeTe—90SnTe	59.9	120 [26]
95GeTe —5TlBiTe$_2$	66.91	555 [25]	SnTe	59.0	75 [27]

TABLE 9. Comparison of Average Heats of Atomization ($\bar{H}_S$, in kcal/g-atom), Average Heats of Formation (Estimated from Electronegativity Differences, Δx: $H_F = 23(\Delta x)^2/7$, in kcal/g-atom) of Defect Tetrahedral Structures (Related to Wurtzite or Zinc Blende Type) and Phenacite-Type Structures

Compound	Structure and type [28]	$\bar{H}_S$	$\bar{H}_F$	$H_S=\bar{H}_S+\bar{H}_F$	Compound	Structure and type [28]	$\bar{H}_S$	$\bar{H}_F$	$H_S=\bar{H}_S+\bar{H}_F$
Zn_3AsI_3	Variation of $HgGa_2Te_4$ type, similar to zinc blende	34.2	2.1	36.3	Zn_2SiO_4	Ph	58.9	9.9	68.8
β-Ag_2HgI_4	HIc type	36.3	1.2	37.5	$LiZnAsO_4$	Ph	56.0	17.5	73.5
Zn_3PI_3	Variation of $HgGa_2Te_4$ type, similar to zinc blende	37.1	2.1	39.2	$LiGaGeO_4$	Ph	62.7	13.1	75.8
β-Cu_2HgI_4	HIc type	39.9	1.2	41.1	$LiZnVO_4^*$	Ph	62.0	14.5	76.5
$AgIn_2Se_3I$	Variation of $HgGa_2Te_4$ type, similar to zinc blende	51.2	3.2	54.4	$LiAlGeO_4$	Ph	63.9	14.5	78.4
In_3AsTe_3	B3	54.4	0.3	54.7	$LiAlSiO_4$	Ph	66.5	14.5	81.0
$CuIn_2Se_3I$	Variation of $HgGa_2Te_4$ type, similar to zinc blende	53.3	3.9	57.2	Be_2GeO_4	Ph	69.6	11.7	81.3
Ga_3AsSe_3	B3	60.6	0.02	60.6	Li_2MoO_4	Ph	68.0	15.9	83.9
Li_2BeF_4	Phenacite (Ph), Si_3 type	36.2	24.8	61.0	Li_2WO_4	Ph	74.0	17.5	91.5
Zn_2GeO_4	Ph	56.3	9.9	66.2	Ge_3N_4	Ph	103.5	4.8	108.3
Li_2SeO_4	Ph	52.5	13.1	65.6	Si_3N_4	Ph	111.3	4.8	116.1
Li_2SO_4	Ph	54.9	13.1	68.0	—	—	—	—	—

* Under pressure, $LiZnVO_4$ transforms from less dense phenacite to more dense spinel type structure [29].

ly it becomes necessary to include the heats of formation in the heats of atomization. The heats of formation are calculated from the corresponding electronegativity differences, and they are added to the average heats of atomization $\bar{H}_S$. The resulting (calculated) values are employed as the heats of atomization.

The compounds in Table 9 are essentially derivatives of the general formula IV_3V_4 (Si_3N_4). They have the zinc blende and wurzite related structures provided their heats of atomization do not exceed approximately 61 kcal/g-atom. When their heats of atomization become greater than this value, they acquire the phenacite structure in which there is a significant deviation from the sp^3-type bonding.

The compounds of Table 10 are of the general formula $II-III_2-VI_4$. When their heats of atomization are less than approximately 64 kcal/g-atom, these compounds have the thiogallate structure whereas when their heats of atomization exceed this value they crystallize in the more dense spinel structure in which the anions are close packed.

Vitreous Semiconductors

In going from the crystalline to the vitreous state there is only a small decrease in the energy gap for a given semiconductor. Such a decrease results from the absence of long-range order (absence of long-range interatomic interactions) in the vitreous state. For example, the energy gap of crystalline As_2Se_3 is 2.1 eV and that of vitreous As_2Se_3 is 2.0 eV [30]. It should be expected that the $\bar{H}_S$ values are directly related to the energy gap values in vitreous just as they are in crystalline semiconductors. Such a correlation is not as yet possible to any extensive degree since the available data on the energy gaps of vitreous semiconductors are very limited.

However, the softening temperature (T_s) of vitreous materials is an approximate measure of their bond strengths. Indeed the correlation between $\bar{H}_S$ and T_s values (together with

TABLE 10. Comparison of Average Heats of Atomization ($\bar{H}_S$, in kcal/g-atom), Average Heats of Formation (Estimated from Electronegativity Differences Δx: $\bar{H}_F = 23(\Delta x)^2/7$, in kcal/g-atom) of Some Compounds with General Formula $II-III_2-VI_4$

Compound	Crystal structure and type [28]	$\bar{H}_S$	$\bar{H}_F$	$H_S = \bar{H}_S + \bar{H}_F$
$HgIn_2Te_4$	Thiogallate (TG), $H1_b$	45.0	0.3	45.3
$CdIn_2Te_4$	TG	46.7	0.5	47.2
$ZnIn_2Te_4$	TG	47.3	0.7	48.0
$HgIn_2Se_4$	TG	47.0	1.2	48.2
$HgGa_2Te_4$	TG	48.2	0.4	48.6
$CdIn_2Se_4$	TG	48.6	1.6	50.2
$CdGa_2Te_4$	TG	49.8	0.7	50.5
$ZnIn_2Se_4^*$	TG	49.3	1.7	51.0
$HgAl_2Te_4$	TG	50.6	0.5	51.1
$ZnGa_2Te_4^*$	TG	50.5	0.8	51.3
$HgGa_2Se_4$	TG	50.1	1.4	51.5
$ZnAl_2Te_4^*$	TG	52.4	1.0	53.4
$CdGa_2Se_4$	TG	51.8	1.7	53.5
$HgAl_2Se_4$	TG	52.5	1.6	54.1
$CdAl_2Te_4^*$	TG	52.3	2.1	54.4
$ZnGa_2Se_4^*$	TG	52.4	2.1	54.5
$CdAl_2Se_4$	TG	54.2	2.1	56.3
$ZnAl_2Se_4$	TG	54.8	2.4	57.2
$HgIn_2S_4$	Spinel (SP)	56.5	1.6	58.1
$CdIn_2S_4$	SP	58.1	2.1	60.2
$ZnIn_2S_4$	SP	58.5	2.4	60.9
$CdGa_2S_4$	SP	61.3	2.4	63.7
$HgAl_2S_4$	SP	62.0	2.1	64.1
$ZnGa_2S_4$	SP	61.9	2.7	64.6
$CdIn_2O_4$	SP	54.9	10.7	65.6
$CdAl_2S_4$	SP	63.7	2.7	66.4
$ZnAl_2S_4^{**}$	SP	64.3	3.0	67.3
$CdGa_2O_4$	SP	59.3	11.3	70.6
$ZnGa_2O_4$	SP	58.1	12.5	70.6
$CdAl_2O_4$	SP	60.5	11.9	72.4
$ZnAl_2O_4$	SP	61.1	12.5	73.6

* For possible alternate tetrahedral structures see Parthé [28]; in compounds such as $ZnGa_2S_4$, it is not possible to make a distinction between alternate structures due to the similar atomic scattering factors of Zn and Ga.

** The low-temperature form for $ZnAl_2S_4$ has the spinel structure, whereas the high-temperature form has a disordered wurtzite type $[(B4)_{H_1}]$ structure [28].

TABLE 11. Comparison of Average Heats of Atomization ($\bar{H}_S$, in kcal/g-atom), Softening Temperatures (T_s, in °C), and Microhardness (mh, in kg/mm^2) of Some Vitreous Materials

Composition	$\bar{H}_S$	T_s [31]	mh [31]	Composition	$\bar{H}_S$	T_s [31]	mh [31]
$As_2Se_3.2Tl_2Se_3$	50.31	96		$As_2Se_3.I_{1,5}$	49.93	85	
$As_2Se_3.Tl_2Se_3$	52.04	109		$As_2Se_3.I_{1,0}$	51.95	108	64
$2As_2Se_3.Tl_2Se_3$	53.77	136		$As_2Se_3.I_{0,5}$	54.35	142	77
$5As_2Se_3.Tl_2Se_3$	55.51	166		As_2Se_3	57.24	187	128
As_2Se_3	57.24	187		$6Tl_2Se.As_2Se_3.Tl_2Te.As_2Te_3$	48.6		
$As_2Se_3.Ge_{0,5}$	60.22	240	128	$Tl_2Te.As_2Te_3$	51.0		61
$As_2Se_3.Ge_{1,0}$	62.7	346	165	$Tl_2Se.As_2Se_3$	52.7		92
$As_2Se_3.Ge_{1,5}$	64.8	356		$Tl_2Se.As_2Se_3.Ge_{0,5}$	54.89		103
$As_2Se_3.Ge_{2,0}$	66.6	425	306	$Tl_2Se.As_2Se_3.Ge_{1,0}$	56.84		114
$As_2Se_3.Ge_{3,0}$	69.53		458	As_2Se_3	57.24	187	128
$As_2Se_3.I_{2,5}$	46.63	35		$Tl_2Se.As_2Se_3.Ge_{2,0}$	60.16		165
$As_2Se_3.I_{2,0}$	48.17	53	54	—	—	—	—

TABLE 12.   Comparison of Average Heats of Atomization
($\bar{H}_S$, in kcal/g-atom), Softening Temperatures (T_s, in °C),
Viscosity (ν, in Stokes at 412°C), Resistivity (ρ, in $\Omega \cdot cm$),
Permittivity (ε, in cgs) and Refractive Index (n) [31]

Composition	$\bar{H}_S$	T_s	ν	ρ	ε	n
As_2Se_3	57.24	187	15.5	$\sim 10^{13}$	12.25	3.50
$4As_2Se_3 . As_2Te_3$	56.83	173	—	$3.6 \cdot 10^9$	14.10	3.76
$3As_2Se_3 . As_2Te_3$	56.73	156	3.88	$1.1 \cdot 10^9$	—	—
$As_2Se_3 . As_2Te_3$	56.22	—	1.23	$1.8 \cdot 10^7$	—	—
$9As_2Se_3 . 11As_2Te_3$	56.12	155	—	—	—	—
$2As_2Se_3 . 3As_2Te_3$	56.02	147	—	$3.3 \cdot 10^6$	18.7	4.34
$As_2Se_3 . 2As_2Te_3$	55.88	—	—	$1.0 \cdot 10^6$	20.0	4.48
$As_2Se_3 . 3As_2Te_3$	55.71	141	—	—	—	—

some other physical characteristics) of various vitreous semiconductor systems shown in Tables 11 and 12 is excellent.

The observed relationship between the average heats of atomization and the physical properties of vitreous materials is obviously very useful in designing new materials with the desired physical properties within the glass-forming region of a system.

Literature Cited

1. H. C. Gatos and A. J. Rosenberg, in: The Physics and Chemistry of Ceramics (ed. by C. Klingsberg), Gordon and Breach, New York (1963), p. 196.
2. E. Mooser and W. B. Pearson, Progr. Semicond., 5:103 (1960).
3. L. Pauling, The Nature of the Chemical Bond, 3rd ed., Cornell University Press, Ithaca, N. Y. (1960).
4. B. F. Ormont, Zh. Neorg. Khim., 5:123 (1960).
5. L. Brewer, in: Electronic Structure and Alloy Chemistry of the Transition Elements (ed. by P. A. Beck), Interscience, New York (1963), p. 222.
6. V. Sadagopan and H. C. Gatos, Solid-State Electron., 8:529 (1965).
7. N. A. Goryunova, The Chemistry of Diamond-like Semiconductors, Chapman and Hall, London (1965).
8. W. F. Schottky and M. B. Bever, Acta. Met., 6:320 (1958).
9. P. Goldfinger, in: Compound Semiconductors (ed. by R. K. Willardson and H. L. Goering), Vol. 1, Reinhold, New York (1962), p. 483.
10. C. Benoit, P. Aigrain and M. Balkanski, Selected Constants Relative to Semiconductors, Pergamon Press, New York (1961).
11. P. Goldfinger and M. Jeunehomme, Trans. Faraday Soc., 59:2851 (1963).
12. S. Misra, Sc. D. Thesis, M.I.T. (1963).
13. G. Busch and M. Moldovanova, Helv. Phys. Acta, 35:500 (1962).
14. G. W. Reimherr, A General Survey of the Semiconductor Field, National Bureau of Standards, Tech. Note 153 (1962).
15. F. Hulliger, Nature, 198:1081 (1963).
16. O. Madelung, Physics of III–V Compounds, Wiley, New York (1964).
17. S. I. Radautsan and N. A. Goryunova, in: Soviet Research in New Semiconductor Materials, Consultants Bureau, New York (1965), p. 1.
18. G. Busch, E. Mooser, and W. B. Pearson, Helv. Phys. Acta, 29:192 (1956).
19. D. R. Mason and D. F. O'Kane, Proc. Fifth Intern. Conf. on Physics of Semiconductors, Prague, 1960, publ. by Academic Press, New York (1961), p. 1026.
20. J. H. Wernick, U. S. Pat. 2,882,467 and 2,882,468 (1959).

21. L. Esaki and P. J. Stiles, private communication.
22. J. P. Suchet, Chemical Physics of Semiconductors, Van Nostrand, New York (1965), p. 98.
23. L. M. Sysoeva, E. Ya. Lev, and N. V. Kolomoets, Fiz. Tverd. Tela, 7:2223 (1965).
24. J. O. Dimmock, I. Melngailis, and A. J. Strauss, private communication.
25. R. Mazeleskey, M. S. Lubell, and W. E. Kramer, J. Chem. Phys., 37:45 (1962).
26. J. W. Bierley, L. Muldawer, and O. Beckman, Acta Met., 11:447 (1963).
27. S. I. Novikova and L. E. Shelimova, Fiz. Tverd. Tela, 7:2544 (1965).
28. E. Parthé, Crystal Chemistry of Tetrahedral Structures, Gordon and Breach, New York (1964).
29. G. Blasse, J. Inorg. Nucl. Chem., 25:136 (1963).
30. A. L. Gubanov, Fiz. Tverd. Tela, 4:1510 (1962).
31. B. T. Kolomiets, Phys. Status Solidi, 7:359, 713 (1964).

HEATS OF ATOMIZATION AND SOME
THERMOCHEMICAL CONSTANTS OF $A^{III}B^{V}$ COMPOUNDS*

L. I. Marina, A. Ya. Nashel'skii, and B. A. Sakharov

An analysis was made of the published experimental and calculated values of the thermochemical constants of gallium phosphide (the standard entropy and the specific heat, the enthalpy, and the free energy of formation of this compound) and of the equilibrium constants and vapor pressure of phosphorus. Approximate methods were used to calculate the remaining constants: the specific heat was found by the Landiya method and the standard entropy was deduced from the Eastman equation and by summing the entropies of elemental group IV semiconductors.

The relative scarcity of information on the thermochemical constants of $A^{III}B^{V}$ compounds is due to their relatively high melting points and the high vapor pressures of the volatile components at these points. Moreover, some of the components of these compounds and the compounds themselves are very aggressive at high temperatures. The present paper gives a systematic analysis of the available data for $A^{III}B^{V}$ compounds and reports some estimates of the missing thermochemical constants. A comparison is made of the available and newly calculated data.

Our approach was based on the method due to Goldfinger [1], who suggested that some of the thermochemical constants of the $A^{III}B^{V}$ compounds could be calculated by summing the corresponding constants of elemental group IV semiconductors. In each case we selected a pair of group IV elements whose atomic weights added up approximately to the molecular weight of the $A^{III}B^{V}$ compound in question.

Combinations of elemental semiconductors suitable for our purpose are listed in Table 1.

Enthalpy and Heat of Atomization

The heat of atomization is an important parameter which is a measure of the chemical bonding in a semiconducting compound and is well correlated with the electrical and the thermochemical properties [2, 3, 23]. The heat of atomization ΔH^{0}_{at} is the sum of the enthalpy of formation of a compound ΔH^{0}_{298} from its component atoms and of the heats of vaporization of the components ΔH^{0}_{g}:

$$\Delta H^{AB}_{at} = \Delta H^{AB}_{298} + \Delta H^{A}_{g} + \Delta H^{B}_{g}. \tag{1}$$

---

*"Crystals," pp. 232-238 (see page 3).

TABLE 1. Combinations of Elemental Group IV Semiconductors Simulating $A^{III}B^V$ Compounds

$A^{III}B^V$ compound		Elemental semiconductors	
formula	molecular weight	elements*	sum of atomic weights
BN	24.8	2C	24.0
AlN	40.98	C + Si	40.1
GaN	83.72	C + Ge	84.6
InN	128.76	C + Sn	130.7
BP	41.82	Si + C	40.1
AlP	57.98	2Si	56.18
GaP	100.7	Si + Ge	100.7
InP	145.7	2Ge	145.2
BAs	85.73	Ge + C	84.6
AlAs	101.9	Ge + Si	100.7
GaAs	144.6	2Ge	145.2
InAs	189.7	Ge + Sn	191.3
BSb	132.0	Sn + C	130.7
AlSb	148.7	Sn + Si	146.8
GaSb	191.4	Sn + Ge	191.3
InSb	236.4	2Sn	237.4

*Sums of the atomic weights of the group IV elements in this column are approximately equal to the molecular weights of $A^{III}B^V$ compounds in the first column.

Table 2 lists the published and the inferred values of the heats of atomization.

There are no published values of the heats of atomization of boron and aluminum phosphides, arsenides, and antimonides. The published enthalpies of formation of the other $A^{III}B^V$ compounds are often highly contradictory. Because of this, we estimated the enthalpies of formation of all $A^{III}B^V$ semiconducting compounds by subtracting the heats of vaporization of the components of a given compound from the heat of atomization of this compound. The latter was found by summing the heats of atomization of the corresponding elemental semiconductors (Table 1). The results of such calculations are presented in Table 3.

The method of calculating the heats of atomization of $A^{III}B^V$ compounds by summing the heats of atomization of the corresponding elemental semiconductors (Table 1) is very simple compared with the experimental determinations of these quantities and its accuracy is quite satisfactory for approximate calculations. The large discrepancies found for aluminum, gallium, and indium nitrides are probably due to seriously underestimated experimental values of the enthalpies of formation of these compounds, which do not agree with the enthalpies of the phosphides, arsenides, and antimonides of these three elements. It is evident from Table 3 that the heats of atomization of $A^{III}B^V$ compounds increase with decreasing molecular weight.

Figure 1 shows the dependences of the heats of atomization of $A^{III}B^V$ compounds on their molecular weights. These dependences are linear for all compounds with the exception of boron

TABLE 2. Heats of Atomization (kcal/mole) of Elements A^{III} and B^V, and of Elemental Group IV Semiconductors

Element	$-\Delta H^0_{298}$	Element	$-\Delta H^0_{298}$
B s	107.5 [4]	As s	79.0*
Al s	75.0 [5]	Sb s	72.0*
Ga s	66.0 [5]	Si s	89.0 [5]
In s	58.2 [5]	Ge s	78.4 [4]
N g	113.7 [6]	Sn s	70.0 [6]
P s	75.2 [5]	C s	138.0 [7]

*Inferred by the present authors.

TABLE 3. Enthalpies of Formation and Heats of Atomization of $A^{III}B^V$ Semiconducting Compounds ($A_s + B_s \rightleftharpoons AB_s$), kcal/mole

Formula	Published values		Values calculated by summing the heats of atomization of elemental semiconductors		Difference, %	
	$-\Delta H^0_{298}$	$-\Delta H_{at}$	$-\Delta H_{at}$	$-\Delta H^0_{298}$	ΔH_{at}	ΔH^0_{298}
BN	60 [8]	281.2	276.0	54.8	−1.8	− 8.6
AlN	63 [8]	251.7	227.0	38.3	−9.8	−39.1
GaN	25 [9]	204.7	216.4	36.7	+5.4	+46.7
InN	—	—	208.0	36.1	—	—
BP	—	—	227.0	44.3	—	—
AlP	—	—	178.0	27.8	—	—
GaP	24.4 [10]	165.6	167.4	26.2	+1.09	+ 7.4
InP	22.3 [11]	155.7	156.8	23.4	+0.7	+ 4.7
BAs	—	—	216.4	29.9	—	—
AlAs	—	—	167.4	13.4	—	—
GaAs	17.7 [11]	162.7	156.8	11.8	−3.6	−33.3
InAs	13.5 [12]	150.7	148.4	11.2	−1.5	−17.1
BSb	—	—	208.0	28.5	—	—
AlSb	—	—	159.0	12.0	—	—
GaSb	9.90 [13]	147.9	148.4	10.4	+0.3	+ 5.0
InSb	8.20 [14]	138.4	140.0	9.8	+1.6	+19.5

compounds. It is interesting to note that all the points representing boron compounds are located about 46 kcal/mole to the right of the lines representing compounds with the same second component. This may be due to the stronger molecular bonds in boron compounds.

Standard Entropy

Very little published information is available on the entropies of $A^{III}B^V$ compounds (this is true also of most of the other thermochemical constants of these compounds). Many of the published entropies have been calculated.

Thurmond and Frosch [15] established that the values of the standard entropies of $A^{III}B^V$ compounds are in very good agreement with the values calculated by summing the entropies of the corresponding elemental group IV semiconductors.

Table 4 lists the data which were used to calculate the entropies S^0_{298} (and their changes ΔS^0_{298}) of $A^{III}B^V$ compounds.

Table 5 gives the values of the entropies of $A^{III}B^V$ compounds taken from the published papers or found by summing the entropies of elemental semiconductors. This table lists also

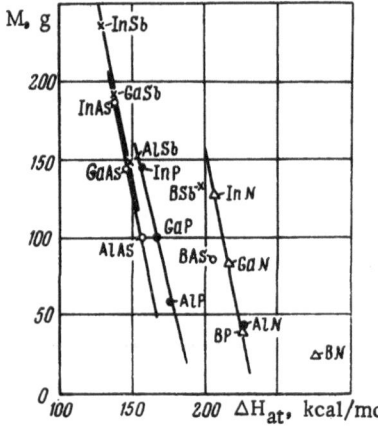

Fig. 1. Dependence of the heats of atomization ΔH_{at} of $A^{III}B^V$ compounds on their molecular weights M.

TABLE 4. Entropies of A^{III} and B^V Elements,
and of Elemental Group IV Semiconductors,
cal · mole^{-1} · deg^{-1} [16]

Element	S^0_{298}	Element	S^0_{298}
B	1.07	As	8.40
Al	6.77	Sb	10.50
Ga	9.80	C	1.36
In	13.98	Si	4.53
N	22.80	Ge	7.43
P	10.60	Sn	10.46

the changes in the entropy calculated from the estimated values for the compounds and from the data given in Table 4.

It is evident from Table 5 that our method for estimating the entropies A^{III}B^V compounds gives values which are in good agreement with the experimental data.

Specific Heat

The experimental values of the specific heat are available only for some A^{III}B^V compounds. Kochetkova and Rezukhina [20] carried out calorimetric measurements of the specific heat of gallium antimonide between 20 and 700°K. The calorimetric method was also used by Nachtrieb and Clement [21] to find the specific heat of gallium antimonide in the temperature range 293-500°K. The temperature dependences of the specific heats of some A^{III}B^V semiconducting compounds were deduced by Piesbergen [19] from the characteristic Debye temperatures. The lack of information on other compounds presents difficulties in the calculations of the thermochemical constants.

In view of this situation, it seemed desirable to obtain at least approximate estimates of the specific heats of A^{III}B^V compounds.

The specific heats of these compounds were calculated by summing the specific heats of the corresponding elemental semiconductors (Table 1).

Table 6 lists the equations for the specific heats of elemental group IV semiconductors, which were used to calculate the specific heats of A^{III}B^V compounds.

TABLE 5. Absolute and Standard Entropies of A^{III}B^V
Compounds, cal · mole^{-1} · deg^{-1}

Compound	S^0_{298}			$-\Delta S^0_{298}$
	published values	calculated values	difference, %	
BN	3.67 [17]	2.72	25.9	21.48
AlN	5.0 [17]	5.89	17.8	23.68
GaN	8.4 [7]	8.79	+4.6	23.89
InN	10.2 [18]	11.82	−15.8	24.88
BP	—	5.89	—	6.11
AlP	—	9.09	—	8.31
GaP	14.39 [19]	11.96	−16.8	8.44
InP	14.28 [19]	14.86	+4.0	9.64
BAs	—	8.79	—	1.01
AlAs	—	11.96	—	3.21
GaAs	15.34 [19]	14.86	3.1	2.34
InAs	18.10 [19]	17.89	−1.7	4.34
BSb	—	11.82	—	0.08
AlSb	15.36 [19]	14.99	−2.4	2.28
GaSb	18.18 [19]	17.89	−1.6	2.41
InSb	20.60 [19]	20.92	+1.5	3.48

TABLE 6. Equations for Specific Heats of
Elemental Group IV Semiconductors [22]

Element	Equation	Temperature range, °K
Sn	$5.05 + 4.80 \cdot 10^{-3}T$	273—504
C	$2.162 + 3.059 \cdot 10^{-3}T - 1.303 \cdot 10^5 T^{-2}$	273—1313
Si	$5.74 + 0.617 \cdot 10^{-3}T - 1.01 \cdot 10^5 T^{-2}$	273—1174
Ge	$5.94 + 0.90 \cdot 10^{-3}T - 0.59 \cdot 10^5 T^{-2}$	273—713

TABLE 7. Specific Heats of $A^{III}B^V$ Semiconducting Compounds

Compound	Coefficients in $C_p = a + bT - cT^{-2}$ (I)			Temperature range, °K	C_p^{298}		Difference
	a	$b \cdot 10^3$	$-c \cdot 10^{-5}$		calculated from Eq.(I)	published values	
BN	4.324	6.118	2.606	273—1313	3.21	4.713 [18]	−1.503
AlN	7.902	3.676	2.313	273—1174	6.396	7.79 [17]	−1.394
GaN	8.102	3.959	1.893	273—713	7.152	—	—
InN	7.212	7.859	1.303	273—504	8.080	—	—
BP	7.902	3.676	2.313	273—1174	6.396	—	—
AlP	11.480	1.234	2.020	273—1174	9.57	—	—
GaP	11.680	1.52	1.60	273—713	10.32	10.4 [10]	−0.08
InP	9.24	4.54	—	273—713	11.08	10.86 [19]	+0.22
BAs	8.102	3.959	1.893	273—713	7.152	—	—
AlAs	11.680	1.520	1.60	273—713	10.320	—	—
GaAs	11.880	1.80	1.18	273—713	11.080	11.05 [19]	+0.03
InAs	10.990	5.70	0.59	273—713	12.026	11.42 [19]	+0.60
BSb	7.212	7.859	1.303	273—504	8.08	—	—
AlSb	10.790	5.417	1.010	273—504	11.26	11.07 [19]	+0.19
GaSb	10.990	5.70	0.590	273—504	12.025	11.60 [20]	+0.42
InSb	10.10	9.60	—	273—504	12.96	12.30 [21]	+0.66

Table 7 gives our estimates of the specific heats of $A^{III}B^V$ compounds. The data presented in this table confirm that the specific heats of these compounds can be calculated by summing the specific heats of elemental group IV semiconductors. Such summation yields results which are in good agreement with the experimental values.

Figure 2 shows that the dependences of the specific heats of $A^{III}B^V$ compounds on their molecular weights are linear for all compounds except those of boron. The points representing boron compounds are located 2.6—2.8 cal · mole^{-1} · deg^{-1} to the left of the straight lines representing compounds with the same second component.

Thus, some of the thermochemical constants (the heat of atomization, the entropy, and the specific heat) can be found by summing the corresponding constants of elemental group IV semiconductors. These elements must be chosen in such a way that the sums of their atomic weights

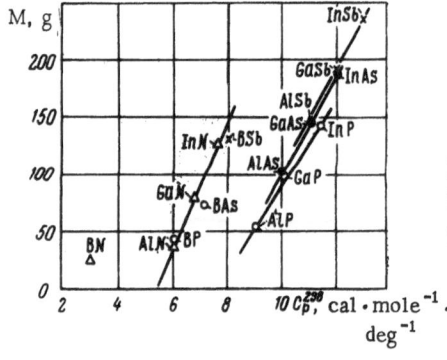

Fig. 2. Dependences of the specific heats C_p^{298} of $A^{III}B^V$ compounds on their molecular weights M.

are nearly equal to the molecular weights of the corresponding compounds. This method is very simple and quite accurate if only approximate thermochemical constants are required. It enabled us to find some of the missing thermochemical constants of $A^{III}B^V$ compounds.

Literature Cited

1. P. Goldfinger, in: Compound Semiconductors (ed. by R. K. Willardson and H. L. Goering), Vol. 1, Preparation of III–V Compounds, Reinhold, New York (1962), p. 483.
2. E. Mooser and W. B. Pearson, Progr. Semicond., 5:103 (1960).
3. W. B. Pearson, Zh. Vses. Khim. Obshchest., 5:493 (1960).
4. F. R. Bichowsky and F. D. Rossini, Thermochemistry of Chemical Substances, New York (1936).
5. Selected Values of Chemical Thermodynamic Properties (ed. by F. D. Rossini, D. G. Wagman, W. H. Evans, S. Levine, and I. Jaffe), Nat. Bur. Stand. (U.S.) Circ. No. 500 (1952).
6. L. Brewer, in: Electronic Structure and Alloy Chemistry of the Transition Elements (ed. by P. A. Beck), Interscience, New York (1963), p. 222.
7. T. L. Cottrell, The Strengths of Chemical Bonds, 2nd ed., Butterworths, London (1958).
8. P. O. Schissel and W. S. Williams, Bull. Amer. Phys. Soc., 4:139 (1959).
9. H. Halm and R. Juza, Z. Anorg. Allg. Chem., 290:82 (1957).
10. C. D. Thurmond, J. Phys. Chem. Solids, 26:785 (1965).
11. B. Stone and D. Hill, Phys. Rev. Lett., 4:282 (1960).
12. T. Renner, Solid-State Electron., 1:39 (1960).
13. W. Schottky and M. Bever, Acta Met., 6:320 (1958).
14. J. Terpilowski and W. Trzebiatowski, Bull. Acad. Polon. Sci., Ser. Sci. Chem., 8:95 (1965).
15. C. D. Thurmond and C. J. Frosch, J. Electrochem. Soc., 111:184 (1964).
16. O. Kubaschewski and E. Ll. Evans, Metallurgical Thermochemistry, 3rd ed., Pergamon Press, London (1958).
17. A. N. Krestovnikov, L. P. Vladimirov, B. S. Gulyanitskii, and A. Ya. Fisher, Handbook for Calculation of Equilibria in Metallurgical Reactions [in Russian], Metallurgizdat, Moscow (1963).
18. M. Kh. Karapet'yants and M. L. Karapet'yants, "Tables of some thermodynamic properties of various substances," Tr. Mosk. Khim.-Tekhnol. Inst., No. 34 (1961).
19. U. Piesbergen, Z. Naturforsch, 18a:141 (1963).
20. N. M. Kochetkova and T. I. Rezukhina, in: Problems in Metallurgy and Physics of Semiconductors (ed. by N. Kh. Abrikosov) [in Russian], Izd. AN SSSR, Moscow (1961), p. 34.
21. N. H. Nachtrieb and N. Clement, J. Phys. Chem., 62:876 (1958).
22. V. G. Ageenkov and Ya. Ya. Mikhin, Metallurgical Calculations [in Russian], Metallurgizdat, Moscow (1962).
23. N. N. Sirota, in: Semiconductors and Semimetals, Vol. 4, Academic Press, New York (1968).

THERMODYNAMIC PROPERTIES OF ALLOYS OF THE $ZnAs_2$–$CdAs_2$ AND Zn_3As_2–Cd_3As_2 QUASIBINARY SYSTEMS*

N. N. Sirota and É. M. Smolyarenko

The emf method was used to determine the thermodynamic functions of the formation of solid solutions in the $ZnAs_2$–$CdAs_2$ and Zn_3As_2–Cd_3As_2 quasibinary systems. Both systems exhibited negative deviations from the Raoult law.

The existence of the $ZnAs_2$–$CdAs_2$ quasibinary alloys was established by Ugai and Zyubina [1]. They reported that solid solutions were formed at all concentrations. The Zn_3As_2–Cd_3As_2 system was investigated by many workers [2–7]. All of them reported a continuous series of solid solutions.

The heats and the entropies of formation of compounds in the Zn–As and Cd–As systems were determined by the emf method and reported in [8, 9]. The present paper deals with the thermodynamic parameters of the formation of solid solutions in the $ZnAs_2$–$CdAs_2$ and Zn_3As_2–Cd_3As_2 systems.

Measurements in which one of the compounds ($ZnAs_2$ or Zn_3As_2) was used as the comparison electrode were unsuccessful because of side reactions in the electrolyte. Therefore, the heats and the entropies of formation of the investigated solid solutions were determined directly by assembling galvanic cells of the type

$$(-)\,Zn_xCd_{1-x}\,|\,KCl + LiCl + ZnCl_2 + CdCl_2\,|\,(Zn_xCd_{1-x})\,As\,(+)$$

and

$$(-)\,Zn_xCd_{1-x}\,|\,KCl + LiCl + ZnCl_2 + CdCl_2\,|\,(Zn_xCd_{1-x})_3\,As_2 + (Zn_uCd_{1-u})\,As_2\,(+).$$

The equations for the reactions of formation of solid solutions in the last galvanic cells were derived for each of the alloys from the positions of the conodes (tie-lines) in the binary regions of the Zn–Cd–As ternary system. We investigated four alloys of the $ZnAs_2$–$CdAs_2$ system and four alloys of the Zn_3As_2–Cd_3As_2 system. The equation for the reaction of formation of any alloy of the $ZnAs_2$–$CdAs_2$ system could be represented in the form $Zn_xCd_{1-x} + 2As = Zn_xCd_{1-x}As_2$. The following equations were used for the Zn_3As_2–Cd_3As_2 system:

$$1)\quad 2Zn_{0,7}Cd_{0,3} + ZnAs_2 = (Zn_{0,8}Cd_{0,2})_3\,As_2,$$

$$2)\quad 2Zn_{0,48}Cd_{0,52} + (Zn_{0,84}Cd_{0,16})\,As_2 = (Zn_{0,6}Cd_{0,4})_3\,As_2,$$

___
* "Crystals," pp. 239–242 (see page 3).

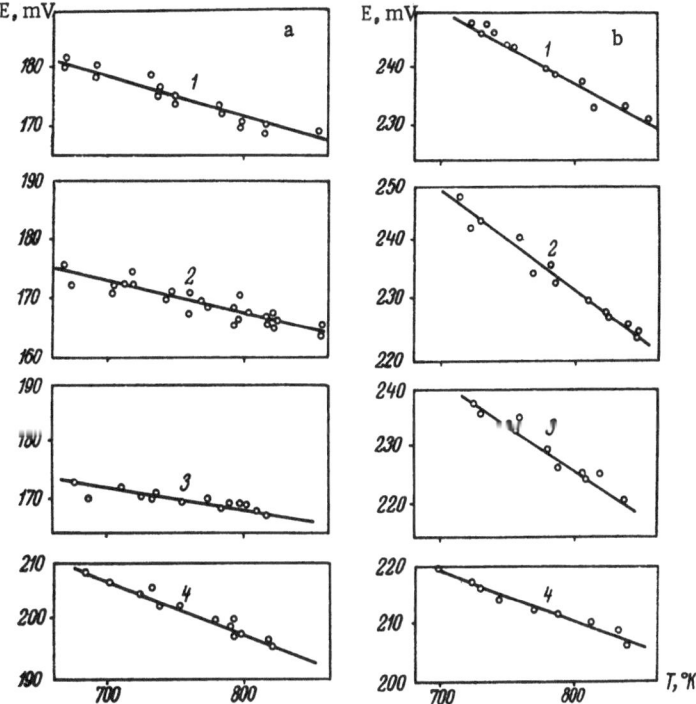

Fig. 1. Temperature dependences of the emf E for: a) $Zn_3As_2 - Cd_3As_2$ alloys (1, 2, 3, and 4 represent, respectively, 0.8, 0.6, 0.4, and 0.2 Zn_3As_2 and 0.2, 0.4, 0.6, and 0.8 of Cd_3As_2); b) $ZnAs_2 - CdAs_2$ alloys (1, 2, 3, and 4 represent, respectively, 0.8, 0.6, 0.4, and 0.2 $ZnAs_2$ and 0.2, 0.4, 0.6, and 0.8 $CdAs_2$).

$$3)\ 2Zn_{0.3}Cd_{0.7} + (Zn_{0.6}Cd_{0.4})\,As_2 = (Zn_{0.4}Cd_{0.6})_3\,As_2,$$

$$4)\ 2Zn_{0.11}Cd_{0.89} + (Zn_{0.38}Cd_{0.62})\,As_2 = (Zn_{0.2}Cd_{0.8})_3\,As_2.$$

The method used to measure the emf was described in our earlier papers [8, 9]. The thermodynamic parameters of the reactions in galvanic cells were deduced from the data presented in Fig. 1. These data were used to calculate the thermodynamic parameters of the formation of the $ZnAs_2-CdAs_2$ and $Zn_3As_2-Cd_3As_2$ solid solutions. In these calculations, we employed the heats and the entropies of formation of the compounds $ZnAs_2$, $CdAs_2$, Zn_3As_2, and Cd_3As_2 found in our earlier investigations. Since the resultant quantities were algebraic sums

TABLE 1. Thermodynamic Functions of Formation of $ZnAs_2 - CdAs_2$ and $Zn_3As_2 - Cd_3As_2$ Solid Solutions

Composition	ΔH, kcal/mole	ΔS, $cal \cdot mole^{-1} \cdot deg^{-1}$	ΔG, kcal/mole	Temp. range, °K
$0.2ZnAs_2 - 0.8CdAs_2$	-2.34 ± 0.9	6.37 ± 1.3	$-2.34 - 6.37 \cdot 10^{-3}T$	697—836
$0.4ZnAs_2 - 0.6CdAs_2$	-3.55 ± 0.86	3.29 ± 1.2	$-3.55 - 3.29 \cdot 10^{-3}T$	704—837
$0.6ZnAs_2 - 0.4CdAs_2$	-2.62 ± 0.94	2.38 ± 1.26	$-2.62 - 2.38 \cdot 10^{-3}T$	697—836
$0.8ZnAs_2 - 0.2CdAs_2$	0.44 ± 1.04	3.32 ± 1.37	$0.44 - 3.32 \cdot 10^{-3}T$	704—854
$0.2Zn_3As_2 - 0.8Cd_3As_2$	-17.9 ± 3.0	-0.57 ± 3.7	$-17.9 + 0.57 \cdot 10^{-3}T$	688—822
$0.4Zn_3As_2 - 0.6Cd_3As_2$	-10.4 ± 2.7	6.8 ± 4.0	$-10.4 - 6.8 \cdot 10^{-3}T$	686—822
$0.6Zn_3As_2 - 0.4Cd_3As_2$	-5.0 ± 2.7	6.4 ± 3.4	$-5.0 - 6.4 \cdot 10^{-3}T$	683—828
$0.8Zn_3As_2 - 0.2Cd_3As_2$	-5.8 ± 1.8	-0.52 ± 2.0	$-5.8 + 0.52 \cdot 10^{-3}T$	670—817

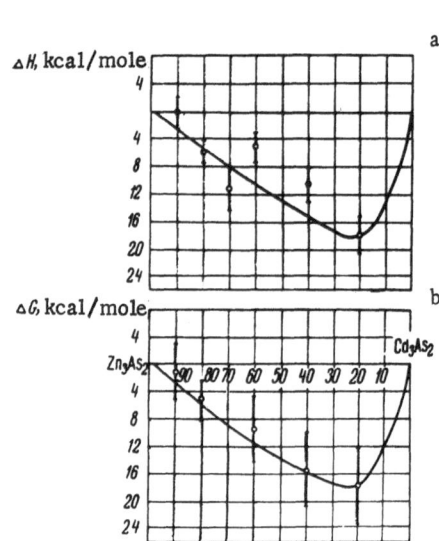

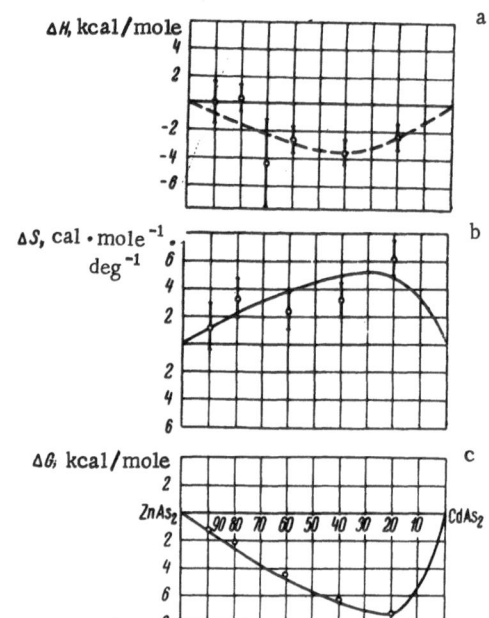

Fig. 2. Dependences of the enthalpy ΔH(a) and of the free energy ΔG(b) of formation of the $Zn_3As_2-Cd_3As_2$ solid solutions on their composition (mol. %).

Fig. 3. Dependences of the thermodynamic functions of formation of the $ZnAs_2-CdAs_2$ solid solutions on their composition (mol. %): a) ΔH; b) ΔS; c) ΔG.

of several terms, the total error was relatively large, particularly in the determinations of the entropy. The results of the calculations are given in Table 1 and in Figs. 2 and 3. It is evident from Figs. 2 and 3 that both systems exhibit negative deviations from the Raoult law. The absolute values of the heats of formation in the $Zn_3As_2-Cd_3As_2$ system are considerably higher than the heats of formation of the $ZnAs_2-CdAs_2$ system. It follows that it is more difficult to form solid solutions in the $ZnAs_2-CdAs_2$ system. This is explained easily by the different crystal structures of the components of this system ($ZnAs_2$ is monoclinic and $CdAs_2$ is tetragonal). The large and positive values of the entropies of formation in the $ZnAs_2-CdAs_2$ system may be also associated with the circumstance that the difference between the structures of the components gives rise to strong lattice distortions and many defects. Solid solutions are formed easily in the $Zn_3As_2-Cd_3As_2$ system because the structure of the two components is the same and the lattice constants differ only slightly.

Literature Cited

1.   Ya. A. Ugai and T. A. Zyubina, Izv. Akad. Nauk SSSR, Neorg. Mater., 2:876 (1966).
2.   A. K. Sreedhar, J. Inst. Telecommun. Eng. New Delhi, 9:248 (1963).
3.   W. Zdanowicz, K. Lukaszewicz, and W. Trzebiatowski, Bull. Acad. Polon. Sci., Ser. Sci. Chim., 12:169 (1964).
4.   H. J. Naake and J. C. Belcher, J. Appl. Phys., 35:3064 (1964).
5.   G. A. Castellion and L. C. Beegle, J. Phys. Chem. Solids, 26:767 (1965).
6.   Ya. A. Ugai, T. A. Zyabina, and E. A. Malygin, Izv. Akad. Nauk SSSR, Neorg. Mater., 2: 17 (1966).

7. N. N. Sirota and É. M. Smolyarenko, Vestsi Akad. Navuk Belarus. SSR, Ser. Fiz. Mat. Navuk, No. 1, p. 107 (1966).
8. N. N. Sirota and É. M. Smolyarenko, in: Chemical Bonds in Semiconductors and Thermodynamics, Consultants Bureau, New York (1968), p. 115.
9. N. N. Sirota and É. M. Smolyarenko, Izv. Akad. Nauk SSSR, Metal., No. 6, p. 234 (1968).

THERMODYNAMIC PROPERTIES OF GALLIUM PHOSPHIDE*

A. S. Abbasov, K. N. Mamedov, and D. M. Suleimanov

An investigation was made of the thermodynamic properties of gallium phosphide using the emf method in the temperature range 320–430°K. The experimental results were used to calculate the free energy, the enthalpy and the entropy of formation, as well as the absolute entropy and the energy of atomization of gallium phosphide at 298°K.

Gallium phosphide (GaP) is one of the $A^{III}B^V$ semiconducting compounds which have the sphalerite structure. The results of some investigations [1, 2] indicate that this material shows promise in the development of high-temperature semiconductor devices.

Gallium phosphide is difficult to investigate because it has a high dissociation vapor pressure and a high melting point. Little is known about the thermodynamic properties of GaP. The published values of the standard heat of formation, deduced by different workers from the dissociation vapor pressure of gallium phosphide, are highly contradictory [3–5].

The thermodynamic properties are needed in the selection of suitable conditions for the synthesis of this compound and of elements which can be used to dope it. Therefore, we determined the principal thermodynamic functions associated with the formation of gallium phosphide. This was done using the emf method and a liquid electrolyte.

We determined, in the 320–430°K range, the emf's of concentrated (relative to the electrodes) electrochemical cells of the type

$$(-)\,\mathrm{Ga_l}\,\left|\begin{smallmatrix}\text{liquid}\\\text{electrolyte}\end{smallmatrix}\right.\!+ \mathrm{GaCl_3}\left|(\mathrm{GaP}+\mathrm{P_{red}})_s\,(+),\right. \tag{I}$$

$$(-)\,\mathrm{Ga_l}\,\left|\begin{smallmatrix}\text{liquid}\\\text{electrolyte}\end{smallmatrix}\right.\!+ \mathrm{GaCl_3}\left|(\mathrm{GaP}+\mathrm{P_{bl}})_s\,(+).\right. \tag{II}$$

The change in the free Gibbs energy (ΔG) as a result of the reaction

$$\mathrm{Ga_l + P_s = GaP_s}$$

was proportional to the emf (E) of the cell:

$$\Delta G = -zFE, \tag{1}$$

where z is the charge of a gallium ion (in our case, $z_{Ga} = 3$ [6, 7]); F is the Faraday number, equal to 23.062 kcal $\cdot$ V^{-1} $\cdot$ (g-eq)$^{-1}$; E is the emf (in volts).

---

* "Crystals," pp. 243–246 (see page 3).

TABLE 1. Free Gibbs Energy, Enthalpy and Entropy
of Formation of 1 Mole of Gallium Phosphide
(T = 320–430°K)

$-\Delta G^0_{375}$, kcal/mole	$-\Delta S^0$, cal · deg^{-1}·mole^{-1}	$-\Delta H^0$, kcal/mole	Reaction
20.0±0.6	5.6±0.9	22.1±1.6	Ga$_1$ + P$_{red}$ = GaP (I)
12.8±0.5	7.1±0.8	15.4±1.4	Ga$_1$ + P$_{bl}$ = GaP (II)

TABLE 2. Standard Heats of Formation and Values of
Absolute Entropy

$-\Delta G^0$, kcal/mole	$-\Delta S^0$, cal deg^{-1}· mole^{-1}	$-\Delta H^0$, kcal/mole	S^0, cal · deg^{-1} · mole^{-1}	Reaction
23.2±0.6	5 4±0.9	24.8±1.6 17.29 [3] 24.4±1.25 [4] 29,1±2.5 [5]	14.3±0.9 9.3 [13]	(I)
22 8±0.5	6.8±0.8	24,9±1,4	12.9±0.8	(II)

A study of the temperature dependence of the emf made it possible to determine the dependence of ΔG on T and thus find the change in the entropy ΔS and enthalpy ΔH resulting from the formation of GaP:

$$\Delta S = -\frac{d(\Delta G)}{dT} = zF\frac{dE}{dT}, \qquad (2)$$

$$\Delta H = \Delta G + T\Delta S = zF\left(T\frac{dE}{dT} - E\right). \qquad (3)$$

The electrodes were mixtures of gallium phosphide (synthesized at the Institute of the Rare–Metal Industry) and red or black phosphors. These mixtures were prepared in an inert atmosphere. Gallium chloride (GaCl$_3$) was prepared by the method described in [4].

The electrolyte was dehydrated and outgassed glycerine containing admixtures of potassium chloride (4 wt.%) and gallium chloride (0.1 wt.%).

The electrolytic cell was an H–shaped Pyrex container suitable for investigation in vacuum. This cell was first introduced and described in detail in [8]. The cell was placed inside an oven and the temperature was maintained to within ± 1–2 deg.

The emf's of the alloys corresponding to the heterogeneous GaP–P region were determined in the 320–430°K range. The method used in these measurements was described in detail in [9, 10]. The temperature dependences of the emf were determined during heating and cooling. The practically identical values of the emf obtained during heating and cooling were evidence of the reversible operation of the electrochemical cell. The experimental values of the emf's obtained for all the electrodes in the heterogeneous region of the GaP–P system were analyzed by the least-squares method [11].

We obtained equations of the E = A + BT type:

$$(I) \quad E = 0.320 - 0.081T \cdot 10^{-3} \pm 0.009 \text{ V}, \qquad (4)$$

$$(II) \quad E = 0.223 - 0.102T \cdot 10^{-3} \pm 0.008 \text{ V}. \qquad (5)$$

The maximum errors in the values of the emf and of the coefficients A and B were deduced from the 95% confidence intervals.

Equations (1)–(5) were used to calculate the free Gibbs energy, and the enthalpy and the entropy of the formation of 1 mole of gallium phosphide from pure liquid gallium and red or black solid phosphorus (Table 1).

Using our and published values of the entropies of the components and the heat of fusion of gallium [12], we calculated the standard values of the thermodynamic functions associated with the formation of gallium phosphide.

Table 2 lists our values of the standard heats of formation, together with published values deduced by various workers from the experimental values of the dissociation pressure and from calculations of the absolute entropy.

The heat of formation reported in [3] was given without an estimate of the experimental error. The entropy given in [13] was calculated using the approximate Eastman formula, which gives values for the $A^{III}B^{V}$ compounds to within ± 5 cal $\cdot$ deg^{-1} $\cdot$ mole^{-1}.

We found that our 298°K values of the standard heats of formation of gallium phosphide from solid gallium and white phosphorus were the same as those obtained using red or black phosphorus. These values were in good agreement with the heat of formation given in [4].

The experimental results were used also to calculate the free Gibbs energy, as well as the entropy and heat of atomization of gallium phosphide. These calculations gave $\Delta H_{298}^{at} = 84.80$ kcal $\cdot$ (g-atom)$^{-1}$, which was close to the mass-spectroscopic value of 85 kcal $\cdot$ (g-atom)$^{-1}$ [14]; $\Delta S_{298}^{at} = 46.95$ cal $\cdot$ deg^{-1} $\cdot$ (g-atom)$^{-1}$; $\Delta G_{298}^{at} = 77.76$ kcal $\cdot$ (g-atom)$^{-1}$.

Literature Cited

1. J. Mandelkorn, Proc. IRE, 47:2012 (1959).
2. J. W. Allen and R. J. Cherry, J. Phys. Chem. Solids, 23:509 (1962).
3. L. I. Marina, A. Ya. Nashel'skii, and S. V. Jakobson, Nauch. Tr. Gos. Nauch.-Issled. Proekt. Inst. Redkometal. Prom., 13:212 (1964).
4. C. D. Thurmond, J. Phys. Chem. Solids, 26:785 (1965).
5. E. N. Ermolenko and N. N. Sirota, in: Chemical Bonds in Semiconductors and Solids, Consultants Bureau, New York (1967), p. 101.
6. H. A. Laitinen, C. V. Liu, and W. S. Ferguson, Anal. Chem., 30:2266 (1958).
7. G. Brauer (ed.), Handbook of Preparative Inorganic Chemistry, 2 vols., Academic Press, New York (1963–1965).
8. A. S. Abbasov, A. V. Nikol'skaya, and Ya. I. Gerasimov, Dokl. Akad. Nauk SSSR, 147:835 (1962).
9. Ya. I. Gerasimov and A. V. Nikol'skaya, Problems in Metallurgy and Physics of Semiconductors [in Russian], Izd. AN SSSR, Moscow (1961), p. 30.
10. A. S. Abbasov, A. V. Nikol'skaya, Ya. I. Gerasimov, and V. P. Vasil'ev, Dokl. Akad. Nauk SSSR, 156:1399 (1964).
11. V. V. Nalimov, Applications of Mathematical Statistics in Analysis of Matter [in Russian], Fizmatgiz, Moscow (1960).
12. D. R. Stull and G. C. Sinke, Thermodynamic Properties of the Elements, Amer. Chem. Soc., Washington (1956).
13. T. Renner, Solid-State Electron., 1:39 (1960).
14. J. Drowart and P. Goldfinger, J. Chim. Phys. Physicochim. Biol., 55:721 (1958).

THERMODYNAMIC PROPERTIES OF
MANGANESE GERMANIDES*

V. N. Eremenko, G. M. Lukashenko, and R. I. Polotskaya

The method of measuring the emf's of galvanic amalgam cells of the type

$$(-) \, \text{Mn}_s \, \left| \, \text{Mn}^{2+}_{\text{ion}} \, \right| \, [\text{Mn} - \text{Ge}]_s \, (+)$$

was used to determine the thermodynamic properties of alloys of the manganese–germanium system in the temperature range 450–700°C. The values of the isobaric-isothermal potentials and of the heats and entropies of formation of the compounds were obtained.

Four metallic compounds — Mn_3Ge_2, Mn_5Ge_3, Mn_5Ge_2, $\text{Mn}_{3.25}\text{Ge}$ — are formed in the Mn–Ge system. One of them, Mn_5Ge_3, has a fairly wide homogeneity region (up to 4 at.% at 600°C) and is capable of dissolving excess manganese (above the stoichiometric ratio). The other three compounds do not exhibit appreciable deviations from their stoichiometric compositions. The solubility of Mn in Ge is very low [1].

Manganese germanides have metallic conduction and can be regarded as semimetals [2]. Three of them (with the exception of $\text{Mn}_{3.25}\text{Ge}$) exhibit ferromagnetic properties [2].

The thermodynamic properties of these compounds have not yet been investigated.

We determined these properties by measuring the emf's of amalgam-type concentrated galvanic cells such as

$$(-) \, \text{Mn}_s \, \left| \, \text{Mn}^{2+}_{\text{ion}} \, \right| \, [\text{Mn} - \text{Ge}]_s \, (+).$$

The principles of the method and the measuring and computation techniques are quite well known [3, 4].

Table 1 gives the phase compositions of the alloys used in our investigation, the chemical reactions responsible for the flow of the current in the galvanic cells, and the equations describing the changes in the isobaric-isothermal potential due to the formation of 1 mole of a given germanide from the pure components.

In our calculations we ignored the homogeneity region of the "Mn_5Ge_3" phase. Our measurements indicated that below 800°C the emf's of the alloys corresponding to this region were close to the emf of cell 3, i.e., the true range of homogeneity of this phase was either narrower

---

* "Crystals," pp. 247–252 (see page 3).

TABLE 1. Alloy Compositions and Current-Generating
Reactions

Cell	Alloy composition	Current-generating reaction	$-\Delta Z^\bullet$
1	$Mn_3Ge_2 + Ge$	$3Mn + 2Ge \to Mn_3Ge_2$	zFE_I
2	$Mn_5Ge_3 + Mn_3Ge_2$	$Mn + 3Mn_3Ge_2 \to 2Mn_5Ge_3$	$\dfrac{zF}{2}(9E_I + E_{II})$
3	$Mn_5Ge_2 + \langle Mn_5Ge_3\rangle$	$5Mn + 2\langle Mn_5Ge_3\rangle \to \to 3Mn_5Ge_2$	$\dfrac{zF}{3}(9E_I + E_{II} + 5E_{III})$
4	$Mn_{3.25}Ge + Mn_5Ge_2$	$1.5Mn + Mn_5Ge_2 \to \to 2Mn_{3.25}Ge$	$\dfrac{zF}{6}(9E_I + E_{II} + + 5E_{III} + 4.5E_{IV})$
5	$\langle \gamma \rangle_{N_{Mn}=0.125} + + Mn_{3.25}Ge$	$Mn + Mn_{3.25}Ge \to \langle \gamma \rangle$	—

or the emf fell very rapidly beyond the stoichiometric composition. Our samples were prepared by fusion in an arc furnace filled with argon. The raw materials were electrolytic manganese ($\geq 99.8\%$) and germanium single crystals of semiconductor purity grade. The samples were subjected to a preliminary homogenizing annealing for 100 h at 650°C. The electrolytes were anhydrous $KCl-LiCl$ and $KCl-NaCl$ mixtures, remelted in vacuum and containing small amounts ($\sim 0.5\%$) of anhydrous $MnCl_2$. All the measurements were carried out in an atmosphere of purified helium.

Figure 1 shows the temperature dependences of the emf's obtained for four cells. The constant value of the emf was established after 5—10 h from the beginning of a run. Subsequently, stable values (within ± 0.2 mV) were reached immediately after the establishment of a constant temperature. The emf's of all the cells did not vary appreciably with time and were easily reproduced during cyclic variation of the temperature. The values obtained were similar for alloys of different composition within the same heterogeneous region, which indicated reversible operation of the galvanic cells. The temperature dependences of the emf's of cells

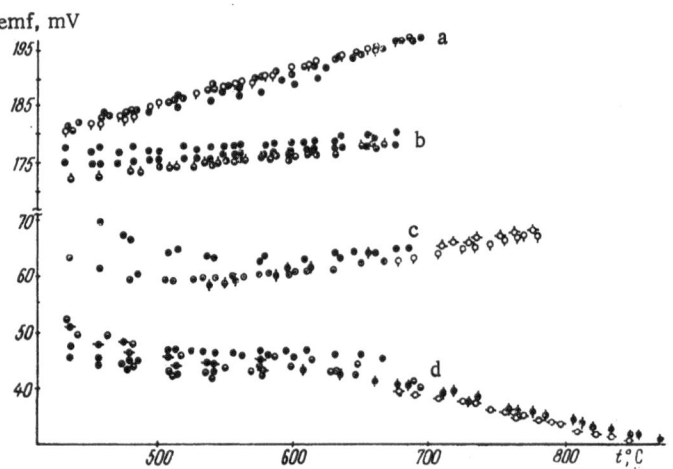

Fig. 1. Temperature dependences of the emf's of galvanic cells: a) 1; b) 2; c) 3; d) 4.

TABLE 2. Equations and Mean-Square Deviations of Experimental Values from Linear Dependences

Cell	emf, V	t, °C	$\pm\overline{\Delta E}\cdot10^{-3}$, V	$-\Delta Z^0$, J/mole	Error, J/mole
1	$0.1451+5.25\cdot10^{-5}T$	440—700	1.1	$83955+30.40T$	±1250
2	$0.1629+1.79\cdot10^{-5}T$	450—680	1.3	$140890+47.33T$	±2000
3	$0.0385+2.62\cdot10^{-5}T$	550—780	1.1	$106145+39.90T$	±2030
4a	$0.0554-1.25\cdot10^{-5}T$	500—650	1.7	$61330+15.87T$	±1450
4b	$0.0940-5.60\cdot10^{-5}T$	560—850	0.9	$66980+11.88T$	±1250
5	$-0.0250+4.10\cdot10^{-5}T$	500—840	0.45	$4985+6.44T$ (J/g-atom)	±350

1 and 2 were practically linear (Fig. 1), whereas those of cells 3, 4, and 5 exhibited a definite nonlinearity at low temperatures. Therefore, we used the emf of cell 3 in the 550—780°C range, where the temperature dependence was nearly linear. The curve obtained for cell 4 was divided into two regions: 1) 500—650°C, and 2) 650—850°C. These two regions were approximated by two linear equations.

The results obtained were analyzed by the least-squares method. We obtained equations of the E = A + BT type. These equations, the mean-square deviations of the experimental values from the linear dependences, and the equations for the calculation of ΔZ^0 are all listed in Table 2.

Table 3 lists the heats and entropies of formation of manganese germanides from α—Mn and solid Ge (these values were the averages obtained at temperatures listed in the second column).

The errors in the estimates of the values obtained were assumed to be twice as large as the mean-square deviations calculated by the least-squares method. These estimates corresponded to the 95% confidence interval for the values within ± 2Δ.

The dependences of the heats and entropies of formation on the concentration in the manganese—germanium system are presented in Fig. 2.

Discussion of the Results

1. The results obtained indicate that the reactions of formation of manganese—germanium compounds are exothermic and the heats of formation of all these compounds are similar. These heats decrease slowly along the Mn_5Ge_3—Mn_3Ge_2—Mn_5Ge_2—$Mn_{3.25}Ge$ series. Thus, the energy of interaction between manganese and germanium does not vary greatly from one compound to another.

TABLE 3. Heats and Entropies of Formation of Manganese Germanides

Compound	Temperature, °C	$-\Delta H^0$, kJ/g-atom	ΔS^0, $J\cdot deg^{-1}\cdot(g\text{-atom})^{-1}$
Mn_3Ge_2	440—700	16.8±0.3	6.1 ±0.36
Mn_5Ge_3	450—680	17.6±0.33	5.95±0.35
Mn_5Ge_2	550—780	15.2±0.47	5.7 ±0.49
$Mn_{3.25}Ge$	500—650	14.4±0.7	3.75±0.5
$Mn_{3.25}Ge$	650—850	15.8±0.52	2.8 ±0.55
⟨γ⟩	500—650	5.0±0.47	6.45±0.30

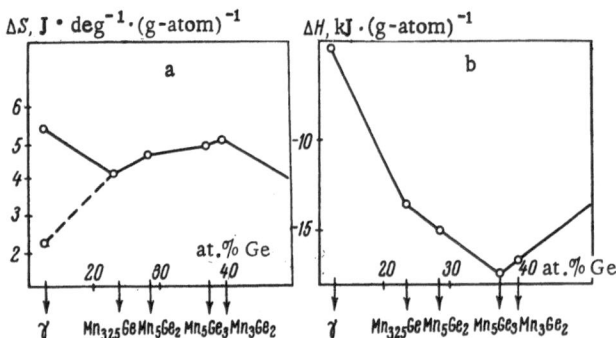

Fig. 2. Entropies (ΔS) and heats (ΔH) of formation
of manganese–germanium alloys.

2. The atomic interaction energies of manganese germanides are approximately 2.5 times smaller than the corresponding energies of manganese silicides [5]. This is a consequence of the weakening of the donor–acceptor interaction resulting from the filling of the 3d band of manganese by the sp electrons provided by Ge or Si [2]. The participation of the d states of manganese in chemical bonding is less in the germanides than in the silicides, and there is a corresponding reduction in the degree of covalence and an enhancement of the metallic nature of the bonds compared with the silicides. The higher silicide, $MnSi_{1.7}$, is a semiconductor, whereas all the compounds of Mn with Ge are characterized by metallic conduction [2].

3. It is interesting to note that the compounds of magnesium with germanium and silicon (Mg_2Si and Mg_2Ge) behave in the opposite manner: the Mg–Ge interaction energy is considerably greater than the Mg–Si energy [4]. This can be explained by the participation of the 3d electrons of germanium in the Mg–Ge chemical bonds, i.e., by the hybridization of the covalent bonds because of the participation of the sp and d states and the corresponding increase in the binding energy. This increase does not occur in the case of Mg_2Si because silicon lacks the 3d electrons (the d states are excited). In the Mn–Si and Mn–Ge compounds, the chemical bonds include a considerable contribution from the d states of manganese and there is no increase in the strength of the binding because of the d states of germanium.

4. In contrast to manganese silicides, whose entropies of formation are large and negative [5], the formation of all the manganese germanides increases the entropy (Table 3). The entropy of the formation of manganese germanides increases somewhat with increasing concentration of manganese (ΔS for the γ solid solution increases because of a configurational contribution to the entropy resulting from a disordered distribution of the atoms: the value of ΔS^{exc} decreases as shown by the dashed line in Fig. 2).

One of the possible reasons for the positive values of ΔS^0 can be indicated in the case of Mn_5Ge_3: the atomic magnetic moment of manganese is higher in the compound than in the pure metallic state [2] and there is a corresponding increase in the magnetic contribution to the entropy.

At temperatures considerably higher than the Curie point (320°K), we can use the standard expression for the magnetic entropy $S_{mag} \approx R \ln(\mu + 1)$, where μ is the atomic magnetic moment, which is 2.57 μ_B [2]. Using this equation and the magnetic entropy of pure α-Mn, which is 4.5 J·deg⁻¹·(g-atom)⁻¹ [6], we obtain

$$\Delta S_{mag} \simeq 3.8 \text{ J} \cdot \text{deg}^{-1} \cdot (\text{g-atom})^{-1}.$$

We can easily see that the magnetic contribution can explain the fairly large value of the total entropy of formation of this compound.

Literature Cited

1.   M. Hansen and K. Anderko, Constitution of Binary Alloys, 2nd ed., McGraw-Hill, New York (1958).
2.   I. G. Fakidov, N. P. Grazhdankina, and V. N. Novogrudskii, Izv. Akad. Nauk SSSR, Ser. Fiz., 20:1509 (1956).
3.   Ya. I. Gerasimov, A. V. Nikol'skaya, V. A. Geiderikh, A. S. Abbasov, and R. A. Vecher, in: Chemical Bonds in Semiconductors and Solids (ed. by N. N. Sirota), Consultants Bureau, New York (1967), p. 87.
4.   V. N. Eremenko and G. M. Lukashenko, in: Chemical Bonds in Semiconductors and Thermodynamics (ed. by N. N. Sirota), Consultants Bureau, New York (1968), p. 103.
5.   V. N. Eremenko, G. M. Lukashenko, and V. R. Sidorko, Porosh. Met., 5:91 (1965).
6.   R. J. Weiss and K. J. Tauer, J. Phys. Chem. Solids, 4:135 (1958).

SOME RELATIONSHIPS GOVERNING
THE DOPING OF SEMICONDUCTORS*

S. A. Semenkovich and Yu. P. Shishkin

Verwey's controlled valence principle is applied to a large number of nontransition metal compounds. This principle makes it possible to predict the sign of conduction of a mixed-valence phase, to explain the appearance of electrical conduction, and to select reliably ways of preparing materials with prescribed thermoelectric properties. The problem of the simultaneous introduction of several impurities into a semiconductor is considered. The role of cation and anion vacancies is determined. The thermo-emf power is estimated from the possibility of the appearance of ions of "abnormal" valence and from the role of the resultant polar bonds. The limits of the validity of Verwey's principle are considered.

The well-known investigations of thermoelectricity carried out by Academician A. F. Ioffe and his school demonstrated the great importance of the introduction of certain impurities into semiconductors. In contrast to the usual doping procedures, Ioffe and his colleagues recommended the use of combinations of chemical compounds and elements ($PbHa_2$ + Pb, Na_2X + X, etc., where Ha is a halogen and X = S, Se, or Te). In view of this, it is necessary to take account of possible exchange reactions, the interaction between solid and gaseous phases, etc.

The knowledge of the chemical nature and the transport properties of impurities is essential in any analysis of doping procedures from the point of view of the optimization of particular properties. The energy band theory does not deal with chemical phenomena and it is of no help in such cases. There have been only a few investigations, mainly experimental, of the appearance of charge carriers as a result of the introduction of complex impurities [38]. However, no attempt has been made to deduce any general relationship. Some workers have suggested fairly complex schemes, which are applicable only to special cases [1–3].

The published work and our own investigations have suggested that Verwey's controlled valence principle [4][1] provides a satisfactory basis for considering the doping of semiconducting compounds.

Verwey's principle has been used in many investigations (for example, [8–17]). However, Verwey and other workers have dealt mainly with transition-metal compounds. We shall show that the principle can be applied to many other substances with variable valence.

---

* "Crystals," pp. 253–265 (see page 3).

[1] This principle was also suggested by Wagner [5] and Anderson [6, 7], independently of Verwey.

Compounds of nontransition metals form solid solutions in relatively narrow ranges of concentration and, consequently they usually contain few ions of "abnormal" valence. However, the number of such ions is still sufficient to alter the thermoelectric properties of such compounds.

It is quite reasonable to assume that bonds of the type considered in [39, 40] should occur in nontransition metal compounds. This applies particularly to the Me–Me' bonds formed between cations of different valences. In such cases, anions can be regarded as having the properties of an inert gas and it is reasonable to assume that they do not make a significant contribution to the total energy of the compound in question.

Without going into details, we may assume that the energy of such bonds is related to the reduction in the free energy observed in the formation of double or more complex salts, oxides, sulfides, tellurides, etc., which has been observed in many substances. The thermodynamic data (ΔS, ΔZ) indicate that this type of bond in double salts becomes weaker with increasing temperature, and this provides an opportunity for electron exchange between increasing numbers of ions having different valences.

Let us consider a lead chalcogenide PbX (X = S, Se, or Te) with an excess of lead or the anion component. This excess provides the necessary condition for the reduction of or increase in the charge of some of the lead cations (the number of these cations is proportional to the excess):

$$PbX + xPb \rightleftarrows Pb_{1-x}^{2+}Pb_{2x}^{+}X^{2-}V_x^X,$$

where V_X^X are the sulfur, selenium, or tellurium vacancies;

$$PbX + xX \rightleftarrows Pb_{1-x}^{2+}Pb_x^{4+}X_{1+x}^{2-}V_x^{Pb},$$

where V_X^{Pb} are the lead vacancies.

The valence of lead in PbS containing excess sulfur, which forms a solution of PbS_2 in PbS, may increase above its normal value [5]. Investigations of the phase diagram of the Pb–Se system [18] have demonstrated the possibility of the formation of $PbSe_2$. Moreover, the existence of tetravalent lead compounds (PbF_4, PbO_2, etc.) demonstrates convincingly that the normally divalent lead cations can have higher valences. The question whether the valence increases to 3+ or 4+ [19] is not of basic importance.

The existence of Pb_2O [20–22] and of lead monohalides [23] shows that a solution in PbX can contain at least some Pb_2X in which lead is monovalent.

The appearance of ions of "abnormal" valences among divalent cations gives rise to electron transitions between ions of different valences. Such transitions (jumps or tunneling) are characterized by low activation energies. In the absence of external directional forces, these transitions occur at random but at a constant average rate.

The presence of a temperature gradient in thermoelectric materials produces a thermoemf. Let us consider what happens in PbX with an excess or a deficiency of cations when this compound is subjected to a temperature gradient. We shall assume that the dissolved Pb_2X or PbX_2 pseudomolecules and the normal PbX molecules are bound by polar bonds: $PbX \cdot Pb_2X$ and $PbX \cdot PbX_2$.

Figure 1 shows schematically a sample of PbX with an excess of Pb (Fig. 1a) or an excess of X (Fig. 1b). At relatively low temperatures, when the conditions for diffusion and chemical reactions are unfavorable, a temperature gradient across a sample does not produce a concentration gradient of cations of different valences. However, the effective concentrations of such cations at the hot and the cold ends of the sample will be different because of the stronger

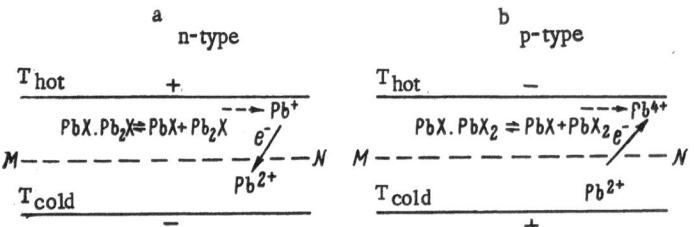

Fig. 1. Schematic representation of the appearance
of a thermo-emf in a sample.

dissociation of the polar bonds in the high-temperature part of the sample. This follows from known chemical observations and from very general thermodynamic considerations.

This difference in the cation concentrations causes directional motion of the charges in a crystal because electrons cross the MN plane predominantly in the direction which results in the leveling of the effective concentrations at the two ends of the sample. In the case represented by Fig. 1a, the electrons moving away from the Pb^+ cations, whose effective concentration is higher in the upper part of the figure, give rise to a preferential motion of the charges across the MN plane in the downward direction. This results in negative charging of the cold part of the sample and positive charging of the hot part, i.e., an n-type thermo-emf is generated.

In samples of PbX with an excess of X, i.e., those containing lead cations of higher-than-normal valence (Fig. 1b), the electrons move across the MN plane mainly toward the Pb^{4+} cations, whose effective concentration is higher in the upper part of the figure. Thus, the electrons move against the temperature field and they charge negatively the hot end of the sample, and positively the cold end. This represents a p-type thermo-emf. Therefore, if a sample of lead sulfide contains cations of normal and lower valences, we should observe n-type conduction. If such a sample contains cations of normal and higher valences, we should have p-type conduction. Otherwise, we would have to assume that, for example, the Pb^{4+} ions lose further electrons, which is in conflict with the energy considerations and the condition of electrical neutrality of the lattice.

These considerations show that, if a given compound contains host cations of a valence lower or higher than the normal value, such a compound can only exhibit one type (n- or p-type) of conduction. Hence, it should be possible to predict a priori the sign of conduction of a mixed valence phase, as noted by Lorenz [24] and extended by us to a larger number of examples (Table 1). The method for predicting the sign of conduction of the mixed valence phase, suggested by Suchet for binary compounds [25] is laborious, requires the knowledge of many parameters, is not always accurate, and is inapplicable to more complex substances.

The sign of conduction in a pure compound determines the shift of the maximum melting point T_{mp} in the homogeneous region away from the strictly stoichiometric composition in the phase diagram.

Our method for predicting the sign of the thermo-emf from the presence of ions of "abnormal" valence can be used also to derive guiding principles for the doping of semiconducting compounds. The introduction of a dopant (an element, compound, or a more complex mixture) results in the formation of some cations of "abnormal" valence.

Table 2 classifies the main cases of doping, ignoring possible side reactions and the possibility that the dopant may be completely insoluble. We shall consider some of the general principles, using lead chalcogenides as our example. It is known that p-type conduction is observed in lead chalcogenides when they are doped with alkali metals [26]. However, the direct

TABLE 1. Predictions of Type of Conduction from Valence
State of Cations

$M_a N_b$	Valence state of M	Other valence states of M	Usual conduction	$M_a N_b$	Valence state of M	Other valence states of M	Usual conduction
CuCl	1	2	p	VN	3(5)	4, 3, 2	n
CuBr	1	2	p	Nb_2O_5	5	4, 3, 2	n
CuI	1	2	p	Ta_2O_5	5	4, 3, 2	n
Cu_2O	1	2	p	MoO_3	6	5, 4, 3, 2	n
Cu_2S	1	2	p	MnO_2	4	3, 2	n
Cu_2Se	1	2	p	$MgFe_2O_4$	3	2	n
Cu_2Te	1	2	p	$NiFe_2O_4$	3	2	n
FeO	2	3	p	$ZnFe_2O_4$	3	2	n
Fe_2O_3	3	2	n	$ZnCo_2O_4$	3	2	n
GaSe	2	1, 3	p	CdS	2	1	n
GaTe	2	1, $\bar{3}$	n	HgS	2	1	n
Ga_2O_3	3	2	n	SnTe	2	4	p
GeS	2	4	p	TiO_2	4	2	n
GeSe	2	4	p	WO_3	6	5, 4, 3, 2	n
GeTe	2	4	p	Ag_2S	1	0	n
In_2O_3	3	2	n	Ag_2Se	1	0	n
MnO	2	3, 4, 6, 7	p	Tl_2O_3	3	1	n
MnS	2	3, 4, 6, 7	p	SiO_2	4	2	n
MnSe	2	3, 4, 6, 7	p	PbO_2	4	2	n
MnTe	2	3, 4, 6, 7	p	VS	2	3, 4, 5	p
NiO	2	3, 4	p	CrS	2	3, 4, 6	p
PbS	2	$\bar{1}$, 4	n	Cr_2O_3	3	4, 5, 6	p
PbSe	2	$\bar{1}$, 4	p	$MgCr_2O_4$	3	4, 5, 6	p
PbTe	2	1, $\bar{4}$	p	$FeCr_2O_4$	3	4, 5, 6	p
SnO_2	4	$\bar{2}$	n	$CoCr_2O_4$	3	4, 5, 6	p
SnS	2	4	p	$ZnCr_2O_4$	3	4, 5, 6	p
SnSe	2	4	p	WO_2	4	5, 6	p
$SnSe_2$	4	2	n	MoS_2	4	5, 6	p
UO_2	4	6	p	Mn_2O_3	3	4, 6, 7	p
V_2O_5	5	4, 3, 2	n	BeS_2	4	4, 6, 7	p
ZnO	2	1	n	FeS	2	3	p
CdO	2	1	n	NiS	2	3	p
SrO	2	1	n	CoO	2	3	p
BeO	2	1	n	PdO	2	3, 4	p
BaO	2	1	n	Tl_2S	1	3	p
CaO	2	1	n	Tl_2O	1	3	p
BaS	2	1	n	GeO	2	4	p
CeO_2	4	3, 2	n	—	—	—	—
ThO_2	4	3, 2	n				
UO_3	6	4, 3	n	Al_2O_3	3	1	n
CdTe	2	1	n	$MgAl_2O_4$	3	1	n
TiS_2	4	3, 2	n	$ZnAl_2O_4$	3	1	n
TiN	3(5)	2	n	PbO	2	4	p
ZrO_2	4	2	n	Sb_2S_3	3	5	p
V_2S_3	3	2	n	Bi_2S_3	3	5	p
$NiAl_2O_4$	2	3	p	$CoAl_2O_4$	2	3	p
SbZn	2	3, 5	p	GaAs	3	1	n
SbCd	2	3, 5	p	InAs	3	2, 1	n
In_2Te_3	3	2, 1	n	SbAl	3	5	p
In_2Se_3	3	2, 1	n	SbGa	3	5	p
In_4S_5	—	3	p	HgSe	2	1	n
$CdSiP_2$	4	2	n	HgTe	2	1	n
$SiMg_2$	4	2	n	Bi_2Te_3	3	$\bar{1}$, 5	n
$SnMg_2$	4	2	n	Sb_2Te_3	3	$\bar{1}$, 5	p
InP	3	2, 1	n	Bi_2Se_3	3	1, $\bar{5}$	n

introduction of alkali metals or their chalcogenides Me^2X (Me = Li, Na, or K) does not produce p-type conduction. This observation has been confirmed many times in investigations at our Institute, as well as in other studies (the examples will be given later).

The doping of PbX with an alkali metal results in the following reaction

$$PbX \quad + \quad 2Me \rightleftarrows Me_2X + Pb,$$

$$\Delta Z \simeq 16-25 \ \frac{kcal}{mole} \qquad \Delta Z \simeq 70-80 \ \frac{kcal}{mole}$$

which has two consequences. The cation–anion ratio of Me_2X is such that, when this compound is dissolved in PbX, anion vacancies are formed in the matrix and this reduces the valence of some of the lead ions in exactly the same way as a deviation from stoichiometry. Secondly, the excess lead resulting from this reaction also reduces the valence of some of the lead ions, i.e., both factors give rise to n-type conduction.

The introduction of Me_2X into PbX generates a proportional number of anion vacancies which are an obstacle to an increase in the valence of lead and, once again, p-type conduction is not obtained. The controlled valence mechanism, which increases the charge of the lead ions, is effective only if the excess sulfur, selenium, or tellurium is not less than that required to form Me_2X_2 with the same cation–anion ratio as the host substance:

$$PbX + 2xMe + 2xX \rightarrow Pb_{1+x}^{2+}Pb_x^{4+}Me_{2x}^+X_{1+2x}^{2-} .$$
$$\text{no vacancies formed}$$

Let us now consider a part of a PbX crystal with an admixture of Me_2X_2, consisting of isolated charged ions. Possible electron transitions in the absence of external forces (these transitions are random but occur at a constant average rate) are indicated by the arrow below:

$$Pb^{2+}X^{2-}Pb^{2+}X^{2-}$$
$$X^{2-}Me^+X^{2-}Me^+$$
$$Pb^{2+}X^{2-}Pb^{4+}X^{2-}$$
$$\nearrow$$
$$X^{2-}Pb^{2+}X^{2-}Pb^{2+}$$

The published examples confirm that it is necessary to introduce an excess of the anion component in doping of this type:

$$\tfrac{1}{2}xLi_2O + (1-x)\,CuO + \tfrac{1}{4}xO_2 \rightarrow Li_xCu_x^{3+}Cu_{1-2x}^{2+}O \ [27],$$

TABLE 2. Principal Cases of Doping of Semiconductors

	Doping of CA		New products in CA	Type of conduction
	element	compound		
Deviation from stoichiometry	1) C 2) A		Soln. C_2A, V_A Soln. CA_2, V_C	n p
Cation of same valence	M	MA	Soln. C_2A, V_A	n No change
Cation of higher valence	M	M_2A_3	Soln. C_2A, V_A	n No change
Cation of lower valence	M	M_2A M_2A+A	Soln. C_2A, V_A Soln. CA_2	n No change p
Anion of same valence	N	CN	Soln. CA_2, V_C	p No change
Anion of higher valence	N	C_mN_2	Soln. CA_2, V_C	p No change
Anion of lower valence	N	CN_2 CN_2+C	Soln. CA_2, V_C Soln. $CN(K_2A)$	p No change n

Note. C and A represent cations and anions of the host substance.

which demonstrates the possibility of the formation of trivalent copper ions responsible for the p-type conduction of CuO. The existence of trivalent copper is reported also in [14, 28]. Other published examples include:

$$x/_2 Li_2S + (1-x) NiS + x/_2S \rightarrow Li_x^+ Ni_x^{3+} Ni_{1-2x}^{2+} S^{2-} \ [29],$$

$$x/_2 Li_2O + (1-x) MnO + x/_4O_2 \text{(air)} \rightarrow Li_x^+ Mn_{(1-2x)}^{2+} Mn_x^{3+} O^{2-} \ [30],$$

$$x/_2 Li_2Se + (1-x) MnSe + x/_2Se \rightarrow Li_x Mn_{(1-x)} Se \ [31].$$

The presence of MnSe and of traces of $MnSe_2$ in $MnSe_{1.01}$ has been deduced from x-ray diffraction data in [31]. Other examples include:

$$x/_2 Li_2Te + (1-x) MnTe + x/_2Te \rightarrow Li_x Mn_{1-x} Te \ [8],$$

$$x/_2 Li_2Te + (1-x-y) MnTe + y MnSe + x/_2Te \rightarrow Li_x Mn_{1-x} Te_{1-y} Se_y \ [32].$$

The heating of $MnTe_2$ produces MnTe [32].

Verwey et al. [4] did not determine the proportion of excess oxygen in the reaction

$$1/_2\delta Li_2O + (1-\delta) NiO + 1/_4\delta O_2 \rightarrow Li_\delta^I Ni_{1-2\delta}^{II} Ni_\delta^{III} O,$$

but simply pointed out that the oxygen is absorbed from the atmosphere. Other workers have determined the amount of oxygen absorbed in this way [33] and have found that this amount agrees with the reaction just given.

The conclusions set out in Table 2 are based on Verwey's mechanism and supported by examples from various sources. We shall consider only those conclusions which are useful in the synthesis of semiconductors. The proportions represented by the cation—anion ratio must be obeyed strictly in the doping of a semiconducting compound and, therefore, an excess of one of the components must be added, in addition to the dopant. The essence of Verwey's mechanism must be understood correctly. For example, an attempt to prepare p-type HgSe by adding pure silver is reported in [34]. When we consider this case, it becomes clear why the expected p-type conduction is not obtained and, in fact, the n-type conduction is enhanced. It follows from the reaction

$$HgSe + 2Ag \rightarrow Ag_2Se + Hg$$

that the addition of silver produces anion vacancies and these vacancies increase the electron density (enhance the n-type conduction).

However, it must be stressed that, in this case, the p-type conduction cannot be obtained even when the anion vacancies resulting by the introduction of silver are compensated by adding an excess of the anion component. This is because Verwey's mechanism would, in this case, imply further increase in the valence of the mercury ions and this is unlikely for energy reasons.

This conclusion is supported by the recent failure to produce p-type HgS [13] by the addition of a monovalent cation and an excess of the anion component of the host substance.

An interesting application of Verwey's mechanism to variable-valence host and impurity ions, resulting in a "blocking" effect, is mentioned by Johnston [14, p. 251]. The essence of this effect is a change of the charge of variable-valence impurity ions when they are added to a compound in addition to a material which can alter the valence of the host cations. This "blocking" effect is manifested by the absence of changes in the thermoelectric properties of a material doped in this way or by the absence of regular changes unless a third variable-valence impurity is added.

Our own experiments on the doping of lead chalcogenides by several impurities introduced simultaneously confirmed this complex manifestation of Verwey's mechanism. The host lattice

TABLE 3. Doping of Lead Chalcogenide (PbX) with Several Impurities*

PbSe, doped with										
Na_2Se_2	Na_3Bi	NaBi	Na_3Sb	NaSb	Na_2Sn	NaSn	"NaAl"	"$NaMg_6$"	"$NaZn_6$"	$NaCd_2$

	Na_2Se_2	Na_3Bi	NaBi	Na_3Sb	NaSb	Na_2Sn	NaSn	"NaAl"	"$NaMg_6$"	"$NaZn_6$"	$NaCd_2$
$\bar{\alpha}$	+38	+53	+84	+52	+115	+70	+75	+42	+43	+37	+42
Hall value of n, cm^{-3}	$8 \cdot 10^{19}$	$5.3 \cdot 10^{19}$	$2.5 \cdot 10^{19}$	$4.9 \cdot 10^{19}$	$1.3 \cdot 10^{19}$	$3.6 \cdot 10^{19}$	$2.7 \cdot 10^{19}$	$7.7 \cdot 10^{19}$	$7.6 \cdot 10^{19}$	$8.6 \cdot 10^{19}$	$7.8 \cdot 10^{19}$

PbTe doped with							
Na_2Te_2	NaAl	Na_3Bi	NaBi	Na_3Sb	NaSb	Na_2Sn	NaSn

	Na_2Te_2	NaAl	Na_3Bi	NaBi	Na_3Sb	NaSb	Na_2Sn	NaSn
$\bar{\alpha}$	+66	+66	+78	−195	+85	−130	+72	+78

*Excess selenium (or tellurium) was added in all cases in amounts necessary to form the corresponding selenide (or telluride). Quotation marks are used for compounds which do not exist in nature. The results presented apply to annealed samples. In all the cases, the amount of Na introduced was $1 \cdot 10^{20}$ cm^{-3}.

of PbX was doped, in addition to Na_2X_2 or Li_2X_2, with chalcogenides of cations whose valence could increase, or which had the highest possible valence: Bi_2X_3, Sb_2X_3, Al_2X_3, SnX, ZnX, CdX, MgX, etc. In all cases when the valence of the impurity cations could not be increased still further, for example, in the case of Al^{3+}, Zn^{2+}, and Mg^{2+}, the number of carriers and the thermoelectric properties were the same as for PbX doped with the same amount of Na_2X_3, and the carrier density was directly proportional to the number of introduced Me^I ions.

In those cases when we added Na_2X_2 as well as ions whose valence could increase (Bi^{3+}, Sb^{3+}, Sn^{2+}), we found that the number of carriers decreased strongly and there was a corresponding change in the thermoelectric properties.

The possibility of the formation, in PbX, of compounds such as Na_3Bi, NaBi, NaSb, etc., had to be rejected because of the thermodynamic considerations indicated by the reactions such as

$$Na + Sb \rightleftarrows NaSb - 15.8 \text{ kcal,}$$
$$2Na + Te \rightleftarrows Na_2Te - 75.0 \text{ kcal,}$$
$$2Sb + 3Te \rightleftarrows Sb_2Te_3 - 28.0 \text{ kcal.}$$

These reactions and the application of the Hess law made it possible to calculate the enthalpy of the reaction in question:

$$Sb_2Te_3 + Na_2Te \rightleftarrows 2NaSb + 4Te + 71.4 \text{ kcal.}$$

The enthalpy $\Delta H \approx +70$ kcal indicated that the reaction could not occur from left to right. We had to reject also the possibility of electron-hole compensation (recombination), postulated by Reiss et al. [35], to explain an increase in the solubility of donors in the presence of acceptors. However, in contrast to the work of Reiss et al., which was concerned with elemental semiconductors, we found that an excess of one of the components had to be introduced into a compound in order to obtain a given type of conduction.

An excess of one of the components restricts the type of conduction that can be obtained and, therefore, the compensation hypothesis has to be rejected. Moreover, it is known that ions such as Sn^{2+} do not increase the carrier density in PbX (Table 3).

It follows that the electron transitions capable of changing the thermoelectric properties are realized most easily between the host cations of different valences. The same conclusions can be reached from Johnston's results [14, p. 250]; he reports that the addition of trivalent iron to NiO does not enhance p-type conduction. It is also supported by the conductivity found in many spinels, where the electron exchange is achieved in the presence of host ions of different valences which may be located at the same crystallographic (tetrahedral or octahedral) positions.

It is interesting to note the well-known relationship between the carrier mobility and the degree of ionicity of a semiconducting compound. There is a definite correlation between an increase in the mobility and a decrease in the ionicity within the same series of compounds [36, 37, 41]. In view of this, we may assume that the ease of exchange of electrons between cations having different valences depends on the nature of the crystal lattice and, particularly, on the distances between the ions. Moreover, this correlation may indicate electron exchange by the tunnel effect because an increase in the covalence strengthens the electron "bridge" between two ions, and this raises the probability of a tunnel transition.

Literature Cited

1. F. A. Kröger and H. J. Vink, Solid State Phys., 3:307 (1956).
2. F. A. Kröger, H. J. Vink, and J. van den Boomgard, Z. Phys. Chem. (Leipzig), 203:1 (1954).
3. F. A. Kröger, The Chemistry of Imperfect Crystals, North-Holland, Amsterdam; Wiley, New York (1964).
4. E. J. W. Verwey, P. W. Haaijman, F. C. Romejn, and G. W. van Osterhout, Philips Res. Rep., 5:173 (1950).
5. C. Wagner, J. Chem. Phys., 18:62 (1950).
6. J. S. Anderson, Proc. Roy. Soc., Ser. A, 185:69 (1946).
7. J. S. Anderson, in: The Physical Chemistry of Metallic Solutions and Intermetallic Compounds, Symposium No. 9, Vol. 2, Paper 7A, HMSO, London (1959).
8. A. J. Panson and W. D. Johnston, J. Inorg. Nucl. Chem., 26:705 (1964).
9. S. G. Tresvyatskii, A. V. Zyrin, and S. A. Maksimenko, in: Chemical Bonds in Semiconductors and Solids (ed. by N. N. Sirota), Consultants Bureau, New York (1967), p. 229.
10. G. G. Mikhailov and V. A. Kozheurov, Zh. Fiz. Khim., 29:775 (1965).
11. A. W. Searcy, Research at High Temperatures [Russian translation], IL, Moscow (1962), p. 332.
12. F. S. Stone, in: Chemistry of the Solid State (ed. by W. E. Garner), Butterworths, London (1955).
13. I. V. Kavich, I. V. Savitskii, and G. I. Il'kiv, Zh. Strukt. Khim., 6:166 (1965).
14. W. D. Johnston, in: Thermoelectricity: Science and Engineering (ed. by R. R. Heikes and R. W. Ure, Jr.), Interscience, New York (1961), pp. 232-284.
15. K. W. Hansen and I. B. Cutler, J. Amer. Ceram. Soc., 49:100 (1966).
16. É. G. Misyuk, O. K. Davtyan, A. N. Sofronkov, and M. V. Uminskii, Elektrokhim., 2:311 (1966).
17. D. N. Poluboyarinov, E. Ya. Shapiro, V. S. Bakunov, and F. A. Akopov, Izv. Akad. Nauk SSSR, Neorg. Mater., 2:336 (1966).
18. A. E. Goldberg and G. R. Mitchell, J. Chem. Phys., 22:220 (1954).
19. Ya. A. Ugai, Introduction to the Chemistry of Semiconductors [in Russian], Vysshaya Shkola, Moscow (1965), p. 206.
20. A. Ferrari, Gazz. Chim. Ital., 56:630 (1926).
21. G. Hägele, Z. Mineral., Geol. Paläont., Ser. A, p. 45 (1937).
22. R. Fricke and P. Ackermann, Z. Phys. Chem. (Leipzig), A161:227 (1932).
23. M. Martiatte, J. Chem. Educ., 26:101 (1949).
24. M. R. Lorentz, Phys. Lett., 12:161 (1964).

25. J. P. Suchet, Chemical Physics of Semiconductors, Van Nostrand, New York (1965).
26. R. W. Fritts, in: Thermoelectric Materials and Devices (ed. by I. B. Cadoff and E. Miller), Reinhold, New York (1960), p. 143.
27. É. G. Misyuk, O. K. Davtyan, and M. V. Uminskii, Elektrokhim., 2:451 (1966).
28. K. Hauffe, Reaktionen in und an Festen Stoffen, 2nd ed., Springer Verlag, Berlin (1966).
29. M. Laffitte and J. Rey, J. Phys. Chem. Solids, 27:277 (1966).
30. W. D. Johnston and R. R. Heikes, J. Amer. Chem. Soc., 78:3255 (1956).
31. W. D. Johnston and R. R. Heikes, J. Amer. Chem. Soc., 80:5904 (1958).
32. A. J. Panson and W. D. Johnston, J. Inorg. Nucl. Chem., 26:701 (1964).
33. Ya. M. Ksendzov, L. N. Ansel'm, L. L. Vasil'eva, and V. M. Latysheva, Fiz. Tverd. Tela, 5:1537 (1963).
34. E. Cruceanu, N. Nistor, and D. Niculescu, Kristallografiya, 11:305 (1966).
35. H. Reiss, C. S. Fuller, and F. J. Morin, Bell Syst. Tech. J., 35:535 (1956).
36. P. Aigrain and M. Balkanski (ed.), Selected Constants Relative to Semiconductors, Pergamon Press, Oxford (1961).
37. H. Welker, Z. Naturforsch., 7a:744 (1952); 8a:248 (1953).
38. T. L. Koval'chik and Yu. P. Maslakovets, Zh. Tekh. Fiz., 26:2417 (1956).
39. W. Klemm, Acti X Congresso Intern. di Chimica, 2:690 (1938).
40. S. M. Ariya and M. M. Khernburg, Zh. Neorg. Khim., 9:1525 (1964).
41. D. A. Wright, Semiconductors, 4th ed., Methuen, London (1966), p. 44.

INVESTIGATION OF THE EVAPORATION OF ANTIMONY AND BISMUTH TELLURIDES AND OF BISMUTH SELENIDE*

Z. Boncheva-Mladenova, A. S. Pashinkin, and A. V. Novoselova

Measurements were made of the saturation vapor pressure of solid antimony and bismuth tellurides and of bismuth selenide. It was found that the evaporation of these compounds was of a dissociative nature, in accordance with the following equations

$$Sb_2Te_{3_s} \rightleftarrows \tfrac{1}{4} Sb_{4_g} + 1\tfrac{1}{2} Te_{2_g}$$

$$Bi_2X_{3_s} \rightleftarrows 2BiX_g + \tfrac{1}{2} X_{2_g}$$

where X = Se, Te. The total vapor pressure satisfied the following equations:

for Sb_2Te_3 $\log P_{mm} = -(10022 \pm 136)/T + 10.929 \pm 0.180$,

for Bi_2Te_3 $\log P_{mm} = -(10443 \pm 444)/T + 11.050 \pm 0.228$,

for Bi_2Se_3 $\log P_{mm} = -(11890 \pm 818)/T + 12.222 \pm 1.036$.

The arms of thermoelectric generators are made of antimony and bismuth tellurides and of bismuth selenide. These compounds are used in their pure or alloyed form. Therefore, investigations of the evaporation of these compounds are of practical interest. A comparison of the differences and the similarities in the evaporation mechanisms and the values of the thermodynamic properties deduced from measurements of the vapor pressures should make it possible to obtain some information on the correlation between the structure of these compounds, on the one hand, and the evaporation mechanism and thermodynamics, on the other. When the present investigation was started (1963) no published data was available on the vapor ressures of antimony and bismuth tellurides and of bismuth selenide. The molecular compositions of the vapors in contact with the antimony–tellurium, the bismuth–tellurium, and the bismuth–selenium systems were investigated by Porter and Spencer [1] by the mass-spectrometric method ($1000-1050°K$). According to Porter and Spencer, the vapor phases contained the following ions: Sb_4^+, Sb_2^+, $SbTe^+$, $Sb_2Te_2^+$, Te_2^+; Bi_2^+, Bi^+, $BiTe^+$, Te_2^+; Bi_2^+, Bi^+, $BiSe^+$, Se_2^+. Porter and Spencer [1] calculated the parameters of the $Me_{2_g} + X_{2_g} = 2MeX_g$ to equilibria at 1000 and 1050°K and employed these parameters to determine the dissociation energies of the gaseous molecules of MeX.

Experimental Investigation

We prepared antimony and bismuth tellurides and bismuth selenide by the fusion of the components in evacuated and sealed ampoules. The antimony, bismuth, and selenium were of the Su–0, V–0, and "rectifier" grades. The tellurium was of 99.99% purity.

---

* "Crystals," pp. 266–276 (see page 3).

The evaporation process was investigated by the Knudsen and the Langmuir methods. The amount of the evaporated substance was deduced from the loss of weight in the evaporation chambers. Special attention was paid to the isolation of a thermocouple from the substance being investigated. This was done by introducing a measuring thermocouple through a tube inserted into the evaporation chamber. This tube had a ground end which was enclosed in a tightly fitting sheath. The thermocouple junction was pressed tightly against the bottom of this sheath. The temperature was maintained and measured to within ± 1 deg. The measuring cells were made of quartz. The effusion apertures in the Knudsen chamber were calibrated by evaporating potassium chloride. The ratio of the area of the effusion aperture to the area of the bottom of the Knudsen chamber was within the range $1/300-1/700$. The experimental results were used to calculate the rate of evaporation ($g \cdot cm^{-2} \cdot sec^{-1}$) and the vapor pressure on the assumption that the substances investigated did not dissociate in the vapor phase.

A comparison of the vapor pressures of all three compounds, determined by the Knudsen and Langmuir methods, showed that their evaporation coefficients differed from unity. Assuming that [2]

$$\alpha = P_L / P_K,$$ (1)

we calculated the values of the evaporation coefficient of the three substances. The evaporation coefficient of bismuth telluride was approximately 0.015 (703–767°K) and that of bismuth selenide was 0.06 (794–900°K). The evaporation coefficient of antimony telluride increased somewhat with falling temperature: its value was 0.13–0.16 (789–705°K).

Thermodynamic Calculations

The results of Porter and Spencer [1] indicate that gaseous molecules of the three chalcogenides investigated (Me_2X_3) do not exist and that the evaporation of these three substances can be described by the following equations:[1]

$$Bi_2X_{3_s} \rightleftarrows 2(1-\alpha) BiX_g + 2\alpha\, Bi_g + \frac{\alpha+1}{2} X_{2_g},$$ (2)

$$Sb_2Te_{3_s} \rightleftarrows 2(1-\alpha) SbTe_g + \frac{\alpha}{2} Sb_{4_g} + \frac{\alpha+1}{2} Te_g.$$ (3)

In order to calculate the vapor pressure on the basis of Eqs. (2) and (3), it is necessary to find a conversion factor K for transforming the relative to the absolute total pressures. This factor may be calculated easily using one of the basic equations for the molecular flow of gases

$$\frac{P_1}{\sqrt{M_1}} = \frac{P_2}{\sqrt{M_2}} = \ldots = \frac{P_n}{\sqrt{M_n}}$$ (4)

and some simple stoichiometric relationships. The derivation of the formula for the calculation of $K = P_{tot}/P'_{rel}$ is given in [3, 4]. It follows from [4] that

$$K = \frac{\sqrt{M_A} \sum_{i=1}^{n} x_i}{\sum_{i=1}^{n} x_i \sqrt{M_i}},$$ (5)

where M_A and M_i are, respectively, the molecular weights of the investigated compound and

---

[1] Under our experimental conditions, the partial pressure P_{Sb_2} was negligibly small and P_{Bi_2} was approximately 2P [5]. Thus, the presence of SB_2 and Bi_2 molecules in the vapor phase could be ignored.

TABLE 1. Standard Heats of Formation of Antimony and Bismuth Chalcogenides

Compound	ΔH_{298}, kcal/mole	Reference	Compound	ΔH_{298}, kcal/mole	Reference
$Sb_2Te_{3_S}$	15.5 ∓ 2.5	[8]	$Bi_2Se_{3_S}$	33.45 ∓ 0.15	[7]
	13.5 ± 0.05	[7]		36.4	[9]
$Bi_2Te_{3_S}$	18.0	[8]	$BiTe_g$	44.7 ∓ 0.3	
	18.75	[7]	$BiSe_g$	39.9 ∓ 0.3	
$SbTe_{g\,as}$	49.0 ∓ 1.3				

one of its vapor components; n is the total number of vapor components; x_i are the stoichiometric coefficients of the vapor components. Thus, using Eq. (2) or (3) and Eq. (5), we can find the total vapor pressure of a chalcogenide representing any degree of dissociation of MeX_g.

The heats of formation of the three chalcogenides being investigated could be measured with high accuracy. This enabled us to estimate the degree of dissociation of diatomic gaseous chalcogenides and thus solve the problem of the mechanism of evaporation of the three chalcogenides.

The published information on the standard heats of formation of the three compounds in question are listed in Table 1. The heats of formation of gaseous diatomic chalcogenides were calculated using the results of Porter and Spencer [1] and the values of the standard heats of sublimation of the corresponding elements, given in [6]. The heats of formation of the Me_2X_3 compounds used in our calculations were taken from [7] (these values were obtained by thermochemical measurements).

The problem of the mechanism of evaporation of the three compounds can be solved as follows. We shall assume that the evaporation takes place at some definite value of the degree of dissociation of a gaseous diatomic chalcogenide. This value will be used to find the total vapor pressure in each measurement. Next, we shall apply the third law of thermodynamics to calculate the heat of dissociative evaporation reaction at 298°K and then the heat of formation of the chalcogenide Me_2X_3. We shall assume that the degree of dissociation of MeX_g remains constant throughout the range covered by our measurements. This method makes it possible to select the dissociative evaporation equations for each of the compounds and then calculate the temperature dependences of the total vapor pressure (the logarithm of the total vapor pressure is deduced from the simple linear dependence on the logarithm of the reaction equilibrium constant). The results of such calculations are presented in Table 2.

TABLE 2. Temperature Dependences of Total Vapor Pressures in Dissociative Evaporation of Antimony and Bismuth Chalcogenides

Dissociative evaporation equation	Temp. range, °K	Times measured	Coefficients in $(\log P_{tot})_{mm} = -A/T + B$	
			$-A$	B
$Sb_2Te_3 \rightleftarrows \frac{1}{2} Sb_{4_g} + \frac{3}{2} Te_{2_g}$ (6)	697–819	12	10022 ± 136	10.929 ± 0.180
$Bi_2Te_3 \rightleftarrows 2BiTe_g + \frac{1}{2} Te_{2_g}$ (7)	718–807	12	10443 ± 444	11.054 ± 0.228
$Bi_2Se_3 \rightleftarrows 2BiSe_g + \frac{1}{2} Te_{2_g}$ (8)	735–879	11	11890 ± 818	12.222 ± 1.036

The values of Φ^{**} for the gaseous selenium and tellurium chalcogenides MeX were taken from [10, 11] and the value for solid Bi_2Te_3 was taken from [12]. We estimated values of Φ^{**} for Bi_2Se_3 and Sb_2Te_3 by empirical methods because no published values were available. Since the reactions of dissociative evaporation of bismuth telluride and selenide were of the same type, we assumed that $\Delta\Phi^{**}$ for reaction (7) was equal to $\Delta\Phi^{**}$ for reaction (8). The value of Φ^{**} of Sb_2Te_3 was calculated as follows. We computed the standard entropy of antimony telluride using the Kopp–Neumann law. The temperature dependence of the specific heat was deduced from the value of C_{p298}, which was also calculated using the Kopp–Neumann law, and from the experimentally determined specific heat, C_{p653} [7].

The enthalpies of reactions (6)–(8) at 298°K were calculated from the third law of thermodynamics and were found to exhibit no systematic temperature dependences.

The same standard enthalpies of reactions (6)–(8) were also calculated from the second law of thermodynamics. We also computed the standard entropies of the solid compounds Me_2X_3. The results obtained are listed in Table 3.

The heats of dissociative evaporation of antimony and bismuth tellurides, calculated using the second and third laws of thermodynamics, are in good agreement. This shows that the suggested dissociative evaporation schemes predominate in the investigated temperature intervals. According to [1], the dissociative process

$$Sb_2Te_{2g} \rightleftarrows 2SbTe_g \qquad (9)$$

is of little importance in the total energy balance. The discrepancy between the heats of dissociative evaporation of bismuth selenide, calculated using the second and third laws of thermodynamics on the basis of Eq. (8), indicate that there is some secondary process in addition to that described by Eq. (8). This process is the thermal dissociation of bismuth selenide:

$$Bi_2Se_{3\,s} \rightleftarrows Bi_l + 1\tfrac{1}{2}Se_{2\,g} . \qquad (10)$$

If the temperature is increased by a few tens of degrees above the values used in our experiments, the condensate resulting from the evaporation of bismuth selenide is found to be rich in selenium. However, the good agreement between the heat of formation ΔH_{298} of Bi_2Se_3 calculated using the third law on the basis of Eq. (8) and that deduced from thermochemical measurements [7] shows that the dissociative evaporation process predominates also in the case of bismuth selenide but is accompanied by the parallel process of thermal dissociation. Antimony and bismuth tellurides can also dissociate thermally but such dissociation is relatively unimportant in the investigated range of temperatures.

We must point out that the standard entropies of formation of Sb_2Te_3 and Bi_2Te_3 calculated from our data are very small. This shows that the two tellurides obey the Kopp–Neumann law.

The dissociative evaporation scheme can also be determined by measuring the total vapor pressure by the Knudsen method and deducing the amount of evaporated matter from the

TABLE 3. Thermodynamic Properties of Antimony and Bismuth Chalcogenides Calculated on the Basis of Assumed Dissociative Evaporation Schemes

Sub-stance	ΔH_{298}^0 of dissociative evaporation, kcal/mole		ΔH_{298}^0 of formation (from third law), kcal/mole	S_{298}^0, cal·deg^{-1}·mole^{-1}		
	from third law	from second law		our finding	from Kopp–Neumann law	from [13]
Sb_2Te_3	86.78±0.07	96.6± 1.2	−12.6±1.6	57.5±1.7	57.5	—
Bi_2Te_3	124.9 ±0.9	124.1± 5	−15.7±1.2	62 2±2.7	62.8	62.4
Bi_2Se_3	130.8 ±0.8	114.7±11.8	−33.9±1.4	—	—	—

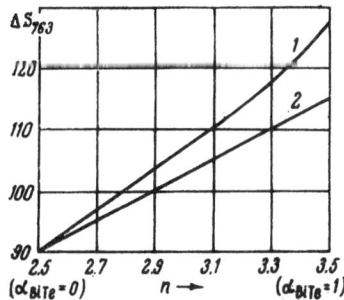

Fig. 1. Graphical calculation of the degree of dissociation of $BiTe_g$ by Gorbov's method [14]: 1) ΔS_{exp}; 2) ΔS_{calc}. Here, n is the number of moles.

loss of weight in the evaporation chamber. In this case, we can use the entropy method developed recently by Gorbov [14]. In this method, the degree of dissociation of gaseous compounds in reactions similar to those represented by Eqs. (2) and (3) is found as follows. We start by calculating the entropies of reaction ΔS_T for various values of α. We then plot the dependences $\Delta S_T = f(\alpha)$. Since the degree of dissociation at a given temperature can have only one value, there is only one point on the curve representing this dependence which corresponds to the real value of α. This point can be found from the intersection with a second dependence $\Delta S_T = f(\alpha)$, which is plotted using the experimental value of P_{tot} at a given temperature, on the assumption that the value of α varies. The total pressure P_{tot} is calculated using Eqs. (5) and (6)−(8). Since this dependence of ΔS_T on α also has one point corresponding to the real value of α, the two curves should intersect at this particular point. A deficiency of Gorbov's method is its low precision because the two curves (lines) intersect at a small angle. Moreover, the relative positions of the curves are affected by the accuracy of the entropies used in the calculations and that of the experimental data. Naturally, the influence of all these factors is very considerable when the curves meet at a small angle.

Our calculations based on Gorbov's method were successful only in the case of the dissociative evaporation of bismuth telluride: we found that the two curves in question intersected at a point corresponding to 2.5 moles ($\alpha_{BiTe} = 0$). Thus, calculations based on Gorbov's method gave the same result as the comparison of the standard heats of formation (Fig. 1). In the case of antimony telluride, the two curves were practically coincident so that it was impossible to determine the point of intersection. The corresponding curves of bismuth selenide were not plotted because we would have had to use the values of ΔS_T deduced from the second law of thermodynamics. Such values are not reliable because (as already mentioned) the slope of the straight line $\log P_{tot} = f(1/T)$ is distorted somewhat by the thermal dissociation of bismuth selenide, which occurs in parallel with the dissociative evaporation.

Discussion of Results

Our investigation shows that the vapor pressures of solid bismuth and antimony tellurides and of bismuth selenide are quite low. The working temperatures of thermoelements made of these substances do not exceed 700°C. Under such conditions, the evaporation of thermoelements should be of little significance, especially as the loss of matter from open surfaces occurs at a rate which is 6−65 times slower than the equilibrium rate of evaporation. The values of the evaporation coefficient (0.15−0.16) found in our study show that the evaporation process is fairly complex. This is supported by thermodynamic calculations, which demonstrate that the evaporation is of a dissociative nature.

The dissociative evaporation of antimony telluride differs from the evaporation of bismuth selenide and telluride. The cause of this difference becomes obvious when we compare the molecular compositions of the metallic antimony and bismuth vapors. The principal component of the antimony vapor at temperatures and pressures corresponding to our measurements is

TABLE 4. Values of ΔG^0_{800} and ΔG_{800} for the Reactions of Formation of MeX

Reaction	ΔG^0_{800}, kcal/mole	ΔG_{800}, kcal/mole
$2Bi_g + Se_{2g} \rightleftarrows 2BiSe_g$ (11)	−27.8	−5.7
$2Bi_g + Te_{2g} \rightleftarrows 2BiTe_g$ (12)	−25.3	−4.9
$\frac{1}{2} Sb_{4g} + Te_{2g} \rightleftarrows 2SbTe_g$ (13)	17.6	9.6

Sb_4. Information on the composition of the bismuth vapor is less definite [5] but it is known that the principal component, under the conditions used in our study, is the monatomic bismuth. This is why the equilibrium in the dissociation of gaseous MeX is shifted in the direction of the formation of binary bismuth selenide and telluride, whereas the equilibrium in the case of $SbTe_g$ is shifted in the direction of dissociation. Thermodynamic calculations confirming this assumption are presented in Table 4. These calculations are based on the data given in [1, 10, 12, 15]. The values of ΔG_T are calculated for the total pressures which correspond approximately to the values of the pressures obtained in our experiments at 800°C.

Our results can be used to compare the thermal stability of bismuth sulfide, selenide, and telluride, whose compositions can be described by Me_2X_3. At temperatures of 700°K and higher, the main process in the case of bismuth sulfide is the thermal dissociation [16−19]

$$Bi_2S_{3\,s} \rightarrow 2Bi_l + \frac{3}{2}\,S_{2\,g} \, , \tag{14}$$

whereas in the case of bismuth selenide and telluride (in the same range of temperatures) the main process is the dissociative evaporation. The possibility of parallel thermal dissociation is not excluded but our data indicate that the contribution of this process to the mass and heat balance is small and that it is more significant in the case of bismuth selenide than in the case of bismuth telluride.[2] Unfortunately, the lack of values of some important thermodynamic properties, particularly in the case of bismuth selenide, and the presence of intermediate phases in the $Bi-Bi_2Se_3$ and $Bi-Bi_2Te_3$ systems, whose thermodynamic functions are not known, make it impossible to calculate the dissociation pressures of bismuth selenide and telluride (Bi_2X_3).

It is quite difficult to explain this difference in the behavior of similar compounds. The factors which can account for these differences in the evaporation mechanisms can be reduced to two. First, we have the presence of intermediate phases which are less rich in Se or Te, and which are not observed in the bismuth−sulfur system [20−22]. The presence of intermediate phases, even though they decompose peritectically below the melting points of Bi_2X_3, should stabilize bismuth selenide and telluride against thermal dissociation. Secondly, bismuth selenide and telluride have one type of crystal structure and bismuth sulfide has a different structure. The telluride and selenide have layered lattices and the sulfide has a chain lattice but with some sulfur atoms outside the chains [23, 24]. It is natural to assume that such sulfur atoms are bound less strongly to the lattice and this accounts for the ease of thermal dissociation in bismuth sulfide.

Finally, we must bear in mind that bismuth and sulfur differ most in their melting points, and this may be related to the predominance of the thermal dissociation in the case of bismuth sulfide.

*   *
*

---

[2] An x-ray diffraction investigation of sublimates of the three bismuth chalcogenides has indicated the presence of considerable amounts of Me_2X_3 compounds, which also supports the proposed evaporation scheme.

Recently, we became aware of Gorbov's paper [14] dealing with the same problems. Therefore, it seemed interesting to compare our results [25] and those reported in [12, 14, 26].

Gorbov also used the Knudsen method and determined the amount of matter lost by evaporation from the change in the weight of a charge in an evaporation chamber. The results of measurements of the vapor pressures of solid antimony and bismuth tellurides were in good agreement with our values. The temperature dependence of the vapor pressure of bismuth selenide obtained by Gorbov [14] and plotted as $\log P = f(1/T)$ was less steep than our dependence. This was due to the fact that Gorbov's samples were slightly richer in bismuth and this could have suppressed the secondary process of the thermal dissociation of bismuth selenide.

The composition of the vapor phase was determined by Gorbov using a special method described earlier in our paper. He assumed that the process of evaporation of the investigated compounds could be described by

$$Me_2X_{3_s} = (1-\alpha)\,Me_2X_{3_g} + 2\alpha MeX_g + \alpha/2X_{2_g}\,. \tag{15}$$

In the case of bismuth telluride, Gorbov found that $\alpha = 1$, in full agreement with our results.

In the case of antimony telluride, Gorbov again assumed $\alpha = 1$ but this resulted in disagreements between the published heats of formation of Sb_2Te_3 and $SbTe_g$ (Table 1) and the heats of reaction of dissociative evaporation. The values of the heat of formation ΔH_{298}^0 of $SbTe_g$ calculated on the basis of the results presented in [1] and [14] differed by about 5 kcal/mole. The dissociative evaporation scheme suggested in the present paper was in full agreement with the thermodynamic functions of all the components and with the stability of the Sb_4 molecule.

Gorbov calculated the degree of dissociation of gaseous Bi_2Se_3 and found that, in the investigated range of temperatures, its value was approximately 0.65–0.70. This result was regarded by Gorbov as preliminary. The existence of gaseous Bi_2Se_3 molecules seems highly unlikely, especially as mass-spectrometric investigations of the composition of the vapors in contact with the Bi–S [27], Bi–Se [1], and Bi–Te [1] systems have given no evidence of the presence of Bi_2X_3 molecules.

Conclusions

1. Measurements were made of the vapor pressures of solid antimony and bismuth tellurides and of solid bismuth selenide.

2. It was found that the evaporation of these compounds was of a dissociative nature and could be described by the equations

$$Sb_2Te_{3_s} \rightleftarrows \frac{1}{2}\,Sb_{4_g} + \frac{3}{2}\,Te_{2_g}\,,$$

$$Bi_2Te_{3_s} \rightleftarrows 2BiX_g + \frac{1}{2}\,X_{2_g}\,.$$

The mechanism of evaporation of antimony telluride differed from the mechanisms found for the other two chalcogenides. This difference was attributed to the high stability of the molecules of the gaseous Sb_4 molecules.

3. The rates of evaporation of these three compounds from open surfaces under nonequilibrium conditions were 6–65 times slower than the equilibrium evaporation rate.

Literature Cited

1.   R. F. Porter and C. W. Spencer, J. Chem. Phys., 32:943 (1960).
2.   A. N. Nesmeyanov and N. É. Khandamirova, Usp. Khim., 28:117 (1959).

3. A. S. Pashinkin, Zh. Fiz. Khim., 40:2611 (1966).
4. A. S. Pashinkin, Zh. Fiz. Khim., 42:1511 (1968).
5. A. N. Nesmeyanov, Vapor Pressure of the Chemical Elements, Elsevier, Amsterdam (1963).
6. Thermal Constants of Substances [in Russian], No. 2, Izd. AN SSSR, Moscow (1966).
7. B. W. Howlett, S. Misra, and M. B. Beyer, Trans. AIME, 230:1367 (1964).
8. Ya. I. Gerasimov and A. V. Nikol'skaya, in: Proc. Fourth Conf. on Semiconducting Materials [in Russian], Izd. AN SSSR, Moscow (1961), p. 30.
9. L. L. Andreeva and A. A. Kuryavtsev, Tr. Mosk. Khim.-Tekhnol. Inst., No. 44, p. 25 (1965).
10. S. I. Gorbov and A. N. Krestovnikov, Zh. Fiz. Khim., 40:940 (1966).
11. D. R. Stull and G. C. Sinke, Thermodynamic Properties of the Elements, American Chemical Society, Washington (1956).
12. S. I. Gorbov and A. N. Krestovnikov, Izv. Akad. Nauk SSSR, Neorg. Mater., 2:1702 (1966).
13. E. S. Itskevich, Zh. Fiz. Khim., 35:1813 (1961).
14. S. I. Gorbov, Author's Abstract of Dissertation [in Russian], Moscow Institute of Steel and Alloys (1966).
15. G. F. Voronin, Zh. Fiz. Khim., 40:1381 (1966).
16. Ya. I. Gerasimov, Zh. Obshch. Khim., 7:1333 (1937).
17. Ya. I. Gerasimov, Zh. Obshch. Khim., 10:1070 (1940).
18. D. Cubicciotti, J. Phys. Chem., 67:118 (1963).
19. G. G. Gospodinov, B. A. Popovkin, A. S. Pashinkin, and A. V. Novoselova, Vestn. Mosk. Univ., Khim., 22 (2):54 (1967).
20. N. Kh. Abrikosov, V. F. Bankina, and K. F. Kharitonovich, Zh. Neorg. Khim., 5:2011 (1960).
21. N. Kh. Abrikosov and V. F. Bankina, Zh. Neorg. Khim., 3:660 (1958).
22. V. K. Nikitina and Yu. K. Lobanova, Izv. Akad. Nauk SSSR, Neorg. Mater., 1:1311 (1965).
23. H. Krebs, Angew. Chem., 70:615 (1958).
24. G. B. Bokii, Introduction to Crystal Chemistry [in Russian], Moscow State University (1954), pp. 421, 427.
25. Z. Boncheva-Mladenova, A. S. Pashinkin, and A. V. Novoselova, Izv. Akad. Nauk SSSR, Neorg. Mater., 2:1542 (1966).
26. S. I. Gorbov and A. N. Krestovnikov, Izv. Akad. Nauk SSSR, Neorg. Mater., 2:1698 (1966).
27. D. Cubicciotti, J. Phys. Chem., 67:1385 (1963).

THERMODYNAMIC PROPERTIES OF
Bi_2Te_3, Bi_2Se_3, Sb_2Te_3, AND Sb_2Se_3 *

S. A. Semenkovich and B. T. Melekh

The liquid-electrolyte emf method was used to determine the values of ΔG_T, ΔH_T, and ΔS_T of Bi_2Te_3, Bi_2Se_3, Sb_2Te_3, and Sb_2Se_3 at temperatures of 503-533, 553-583, 530-560, and 430-485°K, respectively. The published (Se, Te, Bi, Sb, and Bi_2Te_3) and calculated (Bi_2Se_3, Sb_2Se_3, Sb_2Te_3) temperature dependences of the specific heat were employed to calculate the changes in the isobaric-isothermal potential, enthalpy, and entropy associated with the formation of Bi_2Te_3, Bi_2Se_3, Sb_2Te_3, and Sb_2Se_3 from their components. The Karapet'yants–Lauthier comparative calculation method was used to estimate the enthalpies of formation of $BiTe_s$ (−8 kcal/mole), $BiSe_s$ (−14.7 kcal/mole), and $SbTe_s$ (−5.8 kcal/mole).

Trivalent bismuth and antimony tellurides and selenides are the most interesting among the compounds formed in the Bi−Te, Bi−Se, Sb−Te, and Sb−Se systems [1−4]. Solid solutions of these compounds are used widely as thermoelectric materials.

The thermodynamic properties of the higher bismuth and antimony tellurides and selenides have been investigated by many authors but the results are contradictory.

The purpose of our study was to determine the thermodynamic properties of bismuth and antimony tellurides and selenides by the emf method.

We measured the emf's in the galvanic cells

$$(-) Me_{s(l)} \left| \begin{array}{c} \text{liquid} \\ \text{electrolyte} \end{array} + MeCl_3 \right| [Me_2X_{3_s} + X_{s(l)}]\,(+),$$

where Me = Sb, Bi, and X = Se, Te.

The change in the free energy as a result of the reactions[1]

$$Me_{s(l)} + 1.5\,X_{s(l)} \rightleftarrows 0.5\,Me_2X_{3_s}$$

was deduced directly from the equilibrium emf

$$\Delta G = -zFE,$$

where z is the valence of the charge-carrying ions (3+); F is the Faraday number; E is the emf (in volts).

---

* "Crystals," pp. 277−281 (see page 3).

[1] In view of the low solubility of bismuth in selenium (< 1 at.%) at our temperatures [1], we assumed that the postulated reaction produced an emf also when the thermodynamic properties of bismuth selenide were measured.

The temperature dependence of the emf enabled us to estimate entropy and the heat of formation of the higher bismuth and antimony tellurides and selenides. These two quantities were deduced from

$$\Delta S = zF \frac{dE}{dT},$$

$$\Delta H = \Delta G + T\Delta S,$$

where dE/dT is the temperature coefficient of the emf (V/deg).

Our samples were prepared by the fusion of the components in evacuated and sealed quartz ampoules. We used bismuth of the V-000 grade, antimony of the Su-000 grade, and selenium and tellurium containing at least 99.99% of the element. The samples prepared in this way were ground and compacted into cylindrical pellets (5 mm in diameter and 7—9 mm high), which were used as the electrodes. Tungsten wires, embedded in the pellets, were used as the current leads.

The electrolyte in the measurements of the thermodynamic properties of bismuth selenide and telluride and of antimony telluride was the easily melted mixture of anhydrous zinc chloride (analytic purity) with sodium and potassium chlorides (chemical purity grade). The melting point of this mixture was $T_{mp} = 208°C$. The thermodynamic properties of antimony selenide were determined using a mixture of aluminum chloride (distilled twice in vacuum) and sodium chloride (chemical purity grade). The melting point of this mixture was $T_{mp} = 150—155°C$.

The method used in these measurements and the construction of the cell were similar to those employed in [5]. The electrodes and the electrolyte were placed in an H-shaped container enclosed by a stainless-steel cylindrical casing filled with purified argon. The construction of the casing was similar to that employed in [6].

The emf was measured with an R307 potentiometer and an M17/1 galvanometer.

The temperature dependence of the emf was determined during heating and cooling. The experimental results, analyzed by the least-squares method, yielded the temperature dependences of the emf's listed in the third column of Table 1.

The expressions given in Table 1 described, within ± 0.5, ± 1.5, ± 2, and ± 1 mV, the experimental data obtained for all four compounds.

The attempts to measure the emf's at higher temperatures (> 600°K), using a eutectic mixture of potassium and lithium chlorides as the electrolyte, were not successful because of the rapid decomposition of the samples. Such decomposition was due to the formation of light products of the reaction between a tungsten or molybdenum current lead and excess selenium or tellurium.

Our values of ΔG, ΔS, and ΔH are listed in Table 2.

The values given in Table 2 should be attributed to the phases which coexisted with free tellurium and selenium.

TABLE 1. Experimental Data on E = f (T)

Compound	Composition, at.% Se or Te	E = f(T), V	Temperature range, °K
Bi_2Te_3	67,5; 71	$E=0.1507—0.022 \cdot 10^{-3}T$	503—533
Bi_2Se_3	63; 66	$E=0.3390—0.154 \cdot 10^{-3}T$	553—583
Sb_2Te_3	64,55; 68,55	$E=0.1028+0.016 \cdot 10^{-3}T$	530—560
Sb_2Se_3	63,5; 68,5	$E=0.2426—0.031 \cdot 10^{-3}T$	430—485

TABLE 2. Thermodynamic Properties of Higher Bismuth
and Antimony Tellurides and Selenides

Compound	T, °K	$-\Delta G_T$, kcal/mole	ΔS_T, cal · mole^{-1}·deg^{-1}	$-\Delta H_T$, kcal/mole
Bi_2Te_3	503	19.30 ± 0.07	-3 ± 0.9	20.8 ± 0.5
	298	19.80 ± 0.07	-1.4 ± 0.9	20.2 ± 0.5
Bi_2Se_3	573	34.7 ± 0.2	-21.2 ± 6.9	46.9 ± 3.9
	298	36.3 ± 0.2	-1.8 ± 6.9	36.8 ± 3.9
Sb_2Te_3	540	15.4 ± 0.3	$+2.2 \pm 6.4$	14.2 ± 3.5
	298	14.9 ± 0.3	$+2.8 \pm 6.4$	14.0 ± 3.5
Sb_2Se_3	450	31.60 ± 0.15	-4.3 ± 1.1	33.5 ± 0.5
	298	32.20 ± 0.15	-4.0 ± 1.1	33.4 ± 0.5

It follows from Tables 2 and 3 that the thermodynamic properties obtained in our investigation were in good agreement with those deduced using a solution calorimeter (the solvent was liquid bismuth at T = 623°K) [7] and with the values reported in [5] for bismuth and antimony tellurides, deduced by the emf method in the 643–683°K range.

The low value of the enthalpy of formation of bismuth selenide, obtained by combustion calorimetry [8], could be due to some errors committed by the authors in the identification of the products of combustion of bismuth selenide. A repetition of this determination by the method of combustion in a calorimetric bomb, reported in [10], gave a value which was in good agreement with those obtained in [5, 7] and in the present investigation.

The high-temperature thermodynamic data obtained in the present study and in [5] were reduced to T = 298°K using the temperature dependence of the specific heat of bismuth telluride in the 550–28°C range [12], the thermodynamic properties of the elements [11], and the temperature dependences of the specific heat of Bi_2Se_3, Sb_2Se_3, and Sb_2Te_3 deduced in the T_{mp} –25°C range by means of empirical Kelly and Kopp–Neumann laws.

The enthalpies of formation of $BiTe_s$, $SbTe_s$, and $BiSe_s$ were estimated by the Karapet'yants-Lauthier method [13] by comparing ΔH_{298}^0 in the Me_2X_3–MeX series, where Me = Ga, In, Sb, Bi, and X = Se, Te. The values obtained were, respectively, −8.0, −5.8, and −14.7 kcal/mole.

The values of the enthalpies of formation of gallium and indium tellurides and selenides, required in these calculations, were taken from [14–16].

TABLE 3. Published Values of ΔH_{298}^0 and ΔG_{298}^0 of Higher Bismuth and Antimony Tellurides and Selenides

Compound	$-\Delta G_{298}^0$, kcal/mole	$-\Delta H_{298}^0$, kcal/mole	Method
Bi_2Te_3		18.8	High-temperature calorimetry [7]
	19.1	17.9	EMF [5]
		8.0	Approximate calculation [8]
		17.9	Calculation from vapor pressure [9]
Bi_2Se_3		33.5	High-temperature calorimetry [7]
		36.4	Combustion calorimetry in bomb [10]
		13.9	Ditto [8]
Sb_2Te_3		13.5	High-temperature calorimetry [7]
	14.2	13.9	EMF [5]
Sb_2Se_3		30.5	High-temperature calorimetry [7]

Literature Cited

1. M. Hansen and K. Anderko, Constitution of Binary Alloys, 2nd ed., McGraw-Hill, New York (1950).
2. N. Kh. Abrikosov and V. F. Bankina, Zh. Neorg. Khim., 3:659 (1958).
3. N. Kh. Abrikosov, V. F. Bankina, and K. F. Kharitonovich, Zh. Neorg. Khim., 5:2011 (1960).
4. N. Kh. Abrikosov, L. V. Poretskaya, and I. P. Ivanova, Zh. Neorg. Khim., 4:2525 (1959).
5. Ya. I. Gerasimov, A. V. Nikol'skaya, V. A. Geiderikh, A. S. Abbasov, and R. A. Vecher, in: Chemical Bonds in Semiconductors and Solids (ed. by N. N. Sirota), Consultants Bureau, New York (1967), p. 87.
6. V. N. Eremenko and G. M. Lukashenko, Ukr. Khim. Zh., 28:462 (1962).
7. B. N. Howlett, S. Misra, and M. B. Beyer, Trans. AIME, 230:1367 (1964).
8. G. Gattow and A. Schneider, Angew. Chem., 67:306 (1955).
9. Z. Boncheva-Mladenova, A. S. Pashinkin, and A. V. Novoselova, Izv. Akad. Nauk SSSR, Neorg. Mater., 2:1542 (1966).
10. L. L. Andreeva and A. A. Kudryavtsev, Tr. Mosk. Khim.-Tekhnol. Inst., No. 49, p. 25 (1965).
11. D. R. Stull and G. C. Sinke, Thermodynamic Properties of the Elements, American Chemical Society, Washington (1956).
12. G. F. Bolling, J. Chem. Phys., 36:1085 (1962).
13. M. Kh. Karapet'yants, Methods for Comparative Calculation of Physicochemical Properties [in Russian], Nauka, Moscow (1965).
14. P. M. Robinson and M. B. Bever, Trans. AIME, 236:814 (1966).
15. A. S. Abbasov, A. V. Nikol'skaya, Ya. I. Gerasimov, and V. P. Vasil'ev, Dokl. Akad. Nauk SSSR, 156:1140 (1964).
16. H. Hahn and E. Burow, Angew. Chem., 68:382 (1956).

ENTHALPIES OF FORMATION OF
SOME RARE-EARTH SULFIDES*

S. A. Semenkovich, V. M. Sergeeva, and A. D. Finogenov

Measurements of the enthalpies of reactions occurring in the liquid state were used to determine the standard enthalpies of formation ΔH_{298}° of the monosulfides LnS (Ln = La, Ce, Pr, Nd, and Gd) with the NaCl-type structure and sesquisulfides Ln_2S_3 (Ln = La, Ce, Pr, and Nd) with the Th_3P_4-type structure. The values of ΔH_{298}^{0} of LaS, La_2S_3, Ce_2S_3, and Nd_2S_3 were compared with the published values. The values of ΔH_{298}^{0} of LaS, PrS, NdS, GdS, and Pr_2S_3 were obtained for the first time. The ionic component of the chemical bonds in the mono- and sesquisulfides decreased with increasing atomic number of the rare-earth element. This was in agreement with the observation that the difference between the electronegativities of the rare-earth elements and sulfur decreased along the lanthanide series.

There is currently much interest in rare-earth chalcogenides. These compounds have a great variety of properties. Some are semiconductors [1–3], others have metallic conduction [4, 5], and some become superconducting at low temperatures [6, 7]. The compounds represented by LnX and Ln_2X_3 (Ln is a rare-earth lanthanide and X is sulfur, selenium, or tellurium) have been investigated more thoroughly than the other compounds.

Rare-earth monochalcogenides are trivalent in the ground state and have metal-type conduction. These compounds, particularly the monosulfides, are highly stable in the thermal sense [8]. They melt without decomposing. For example, the vapor of lanthanum monosulfide consists mainly of LaS molecules [9]. Europium, ytterbium, and samarium monochalcogenides are semiconductors. All the monochalcogenides have the NaCl-type structure.

Rare-earth sesquichalcogenides are insulators at room temperature but at higher temperatures their resistivities fall in the same way as those of semiconductors. These compounds have lower melting points than the monochalcogenides and they decompose on melting. Many sesquichalcogenides have the defect structure of the Th_3P_4 type [10].

Qualitative information on the nature of chemical bonds in rare-earth chalcogenides is given in [3, 11—14].

Thermodynamic data are one of the sources of information on the strength of chemical bonds. This is particularly true of the enthalpies of the formation of compounds. Unfortunately, no thermodynamic data are available for most of the rare-earth mono- and sesquichalcogenides.

The present paper describes a determination of the enthalpies of formation of some rare-earth sulfides with the formulas LnS and Ln_2S_3 (Ln = La, Ce, Pr, Nd, and Gd). The samples of LnS and Ln_2S_3 were fused polycrystalline aggregates. A phase x-ray structure analysis, carried

---

* "Crystals," pp. 282-291 (see page 3).

out in the laboratory of A. I. Zaslavskii, showed that the monosulfides (LnS) which we investigated had well-developed structures of the NaCl type and that the sesquisulfides (Ln_2S_3) had the Th_3P_4-type structure.

Method Used in the Measurement of Enthalpies of Formation

The high solubility of rare-earth sulfides in cold aqueous solutions of inorganic acids [8, 15–19] made it possible to investigate synthesized substances and to determine their compositions by chemical and x-ray analyses.

We dissolved a charge representing 0.01 g-atom of Ln, or 0.01 mole of LnS, or 0.01 mole of $LnS_{1.5}$. Such a charge was dissolved in a 1/20 aqueous solution of HCl ($d_{293} = 1043.4$ kg/m^3). The interaction between rare-earth sesquisulfides and hydrochloric acid solutions should produce hydrogen sulfide [15–17, 19], whereas the interaction of monosulfides should produce hydrogen sulfide and hydrogen [18, 19]. The heat of solution of $H_2S_{(g)}$ in liquid products of these reactions was not known. The absorption of $H_2S_{(g)}$ in these products was avoided by a preliminary saturation of the solution of HCl with hydrogen sulfide [20, 21]. The enthalpy of formation of a rare-earth sulfide could be calculated if we knew the enthalpies of the following reactions:

$$(1) \qquad Ln_{(s)} + 3HCl_{(soln\ HCl\cdot0.031H_2S\cdot20H_2O)} + 24.57HCl\cdot0.765H_2S\cdot491.4H_2O =$$

$$= LnCl_{3\,(soln\ LnCl_3\cdot24.57HCl\cdot0.858H_2S\cdot551.4H_2O)} + 1.5H_{2\,(g)}, \qquad \Delta H_1;$$

$$(2) \qquad LnS_{(s)} + 3HCl_{(soln\ HCl\cdot0.031H_2S\cdot20H_2O)} + 24.57HCl\cdot0.765H_2S\cdot491.4H_2O =$$

$$= LnCl_{3\,(soln\ LnCl_3\cdot24.57HCl\cdot0.858H_2S\cdot551.4H_2O)} + 0.5H_{2\,(g)} + H_2S_{(g)}, \qquad \Delta H_2;$$

$$(3) \qquad LnS_{1.5(s)} + 3HCl_{(soln\ HCl\cdot0.031H_2S\cdot20H_2O)} + 24.57HCl\cdot0.765H_2S\cdot491.4H_2O =$$

$$= LnCl_{3\,(soln\,LnCl_3\cdot24.57HCl\cdot0.858H_2S\cdot551.4H_2O)} + 1.5H_2S_{(g)}, \qquad \Delta H_3;$$

$$(4) \qquad H_{2\,(g)} + S_{(rhomb)} = H_2S_{(g)}, \qquad \Delta H_4.$$

Preliminary experiments showed that the change in the enthalpy ΔH_1 in reaction (1) was practically identical, within the limits of the experimental error, with the change in the enthalpy in the following reaction:

$$Ln_{(s)} + 3HCl_{(soln\ HCl\cdot20H_2O)} + 24.57HCl.491.4H_2O =$$

$$(5) \qquad = LnCl_{3\,(soln\ LnCl_3\cdot24.57HCl.551.4H_2O)} + 1.5H_{2\,(g)}, \qquad \Delta H_5.$$

Reaction (5) was much easier to carry out than reaction (1). Therefore, we determined experimentally the values of ΔH_2, ΔH_3, and ΔH_5. The fourth value was taken from the published data:

$$\Delta H_4 = 20.1 \pm 0.6 \ (kJ/mole) \quad (4.8 \pm 0.15 \ kcal/mole) \qquad [22, 23].$$

The relationship between the joule and the calorie was assumed to be 4.184.

The enthalpies of formation of rare-earth mono- and sesquisulfides

$$(6) \qquad Ln_{(s)} + S_{(rhomb)} = LnS_{(s)}, \qquad \Delta H^0_{298}(LnS),$$

$$(7) \qquad 2Ln_{(s)} + 3S_{(rhomb)} = Ln_2S_{3(s)}, \qquad \Delta H^0_{298}(Ln_2S_3)$$

were determined by the simultaneous solution of Eqs. (2), (4), and (5) and Eqs. (3), (4) and (5). We thus found that

$$\Delta H^0_{298}(LnS) = \Delta H_5 - \Delta H_2 + \Delta H_4, \tag{8}$$

$$\Delta H^0_{298}(Ln_2S_3) = 2\Delta H_5 - 2\Delta H_3 + 3\Delta H_4. \tag{9}$$

Calorimeter

The enthalpies of reactions (2), (3), and (5) were determined in a variable temperature calorimeter with an isothermal jacket (Fig. 1). The calorimeter was constructed in accordance with the recommendations given in [24, 26]. A cylindrical calorimeter casing was made of 0.2-mm-thick nickel sheet. The calorimeter liquid was 2 kg of distilled water. The evaporation of the water was prevented by pouring 8.4 g of vaseline oil on its surface.

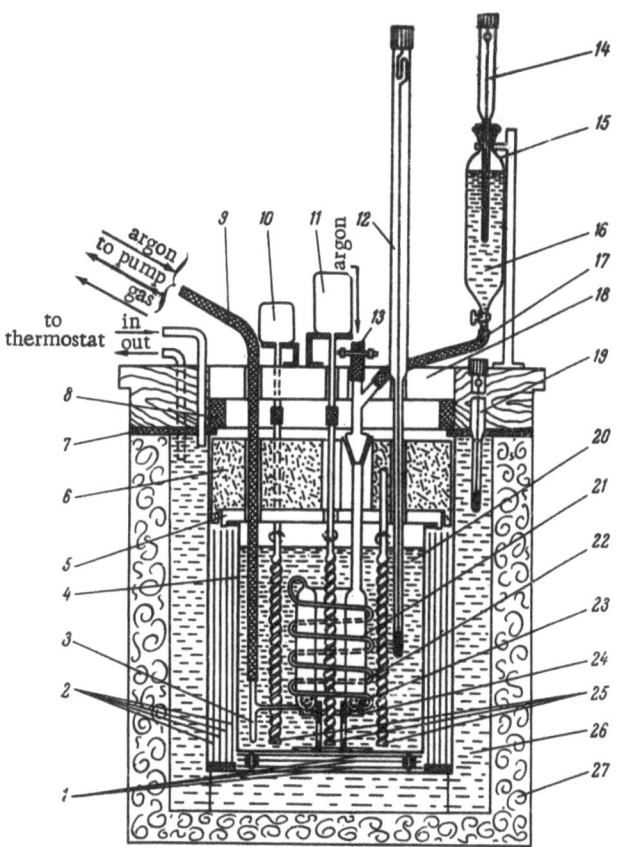

Fig. 1. Construction of the calorimeter: 1) bottom screens; 2) side screens; 3) tube for condensed water vapor; 4) calorimeter casing; 5) calorimeter cover; 6) plastic-foam insulation; 7) rubber spacer; 8) ebonite ring; 9), 13), 17) vacuum rubber tubing; 10), 11) electric motors; 12) Beckmann thermometer; 14), 19) mercury-in-glass thermometers; 15) separatory funnel; 16) solvent; 18) ebonite cover; 20) calorimetric liquid; 21) calorimeter chamber; 22) coil for cooling outgoing gases; 23) investigated substance; 24) ebonite support; 25) glass screw-type stirrers; 26) thermostatting liquid; 27) felt insulation.

The true temperature rise Δt was deduced from the formula

(10) $$\Delta t = \Delta t' + \delta,$$

where $\Delta t'$ is the measured temperature rise; δ is the correction for heat exchange with the ambient medium (δ was calculated using the Regnault–Pfaundler–Usov formula). The value of $\Delta t'$ was measured with a Beckmann thermometer (corrections were made in accordance with the nominal parameters of the thermometer).

The average temperature of the calorimetric liquid during a reaction was 298°K.

The reaction between the given substance and the solvent took place in a glass calorimeter chamber. A charge of the substance being investigated was first placed in the chamber. An inert gas (argon) filling the chamber prevented oxidation of the original substances and of the reaction products. The solvent was placed in a separatory funnel outside the calorimeter. The main measurements were carried out from the moment when the solvent entered the calorimeter chamber from the funnel. The dissolution of Ln, LnS, and Ln_2S_3 was slowed down by covering large parts of the samples with paraffin. A Tishchenko flask, with axle oil as the buffer liquid, separated the calorimeter chamber from the gas-intake system. This system consisted of burettes for measuring the volume of the gas evolved as a result of reactions (2), (3), and (5), and Drexel bottles with an alkali solution of absorbing $H_2S_{(g)}$.

The calorimeter was calibrated by passing an electric current, in accordance with the procedure recommended in [24, 25]. A detailed description of the potentiometric circuit used in the calibration was given in [27].

Solvent

The solvent was prepared from hydrochloric acid of chemical purity grade. The specific heat of the $HCl \cdot 20H_2O$ solution was 3.51 ± 0.03 kJ $\cdot$ kg^{-1} $\cdot$ deg^{-1} [27]. The specific heat of the liquid products of reaction (5) was the same — within the limits of the experimental error — for any one of the rare-earth metals (La, Ce, Pr, Nd, Gd): its value was 3.46 ± 0.03 kJ $\cdot$ kg^{-1} $\cdot$ deg^{-1}.

The specific heat of water, required in the calculations concerned with calibration, was calculated using the formula suggested in [28].

The $HCl \cdot 20H_2O$ solution was placed in a Drexel bottle just before reaction (2) or (3). Hydrogen sulfide was then generated by the action of an aqueous solution of HCl on FeS in Kipp's apparatus, was purified and dried to remove H_2O vapor [29]. Hydrogen sulfide was then passed through the solution. The excess H_2S at the exit from the Drexel bottle was collected in a glass gas jar. The degree of saturation of the $HCl \cdot 20H_2O$ solution with hydrogen sulfide was determined by iodometric back titration of a series of samples. The 0.1 N solutions of iodine and sodium thiosulfate were prepared and standardized in accordance with [30]. The concentration of H_2S in a saturated $HCl \cdot 20H_2O$ solution was 1.4 times smaller than the equilibrium concentration of H_2S in water at 298°K.

The specific heat of the H_2S-saturated acid solution ($HCl \cdot 0.031\ H_2S \cdot 20H_2O$) was 3.50 ± 0.03 kJ $\cdot$ kg^{-1} $\cdot$ deg^{-1}. The concentration of H_2S in the liquid products of reactions (2) and (3) was the same as that before the reaction (within the limits of the precision of the iodometric titration). The specific heat of the liquid products of reactions (2) and (3) was 3.45 ± 0.03 kJ $\cdot$ kg^{-1} $\cdot$ deg^{-1}.

Synthesis of Sulfides

The LnS and Ln_2S_3 sulfides were synthesized from their components. The purity of the rare-earth metals is illustrated by the following data for lanthanum. This metal was of 0 grade and was prepared by electron-beam melting. No Ce or Ca was found in this metal (to within 0.01%) and there was no Nd (to within 0.005%). The concentrations of other impurities were: 0.005–0.009% Pr, 0.008–0.017% Fe, 0.005–0.008% Cu, 0.1–0.13% O_2. The other rare-earth metals were of similar quality. The concentration of the main component was 99.5–99.7%.

The sulfur used in the synthesis was of the V-4 grade. A detailed description of the synthesis method was given in [31].

Mono- and sesquisulfides were synthesized by the reaction between rare-earth metals and sulfur vapor. A fine powder of uniform grain size obtained in this way was compacted, annealed at a high temperature, and then melted in an induction furnace. A crucible containing a sample was placed inside a quartz ampoule filled with argon. The monosulfides were melted in tantalum crucibles and the sesquisulfides in graphite crucibles. The ampoule was lowered through the induction-heated zone at a rate of 80–20 mm/h. In the case of the sesquisulfides, the ampoule was heated additionally to 800–900°C (resistance heaters were placed at the top and the bottom of the induction furnace). The temperature was measured with an optical pyrometer. The monosulfides were obtained in the form of coarse-grained ingots of a golden-yellow color; in some cases, single crystals were obtained.

The chemical analysis confirmed, within the limits of the experimental error, that our compounds were stoichiometric. This analysis was carried out using the method described in [32].

We measured the microhardness of the monosulfides using a PMT-3 test unit calibrated against NaCl. The surfaces of the samples were single-crystal cleavage planes of the (100) type. The diagonals of the indentations were within the limits 20–25 μ for a load of 50 g. The following values of the microhardness (kgf/mm^2) were obtained: 150 (LaS), 180(CeS), 190 (PrS), 220 (NdS), and 290 (GdS).

Experimental Results and Discussion

The values of ΔH_2, ΔH_3, and ΔH_5 were determined by carrying out three measurements for each quantity. The experimental results were analyzed to give arithmetic means of the enthalpy.

Table 1 lists the experimental values of the enthalpies of reactions (5), (2), and (3), which take account of the experimental errors and the reproducibility of the measurements. These values of ΔH_5, ΔH_2, and ΔH_3 were used to calculate the enthalpies of formation of the rare-earth mono- and sesquisulfides in accordance with Eqs. (8) and (9). The results are given in Table 2. This table includes also the experimental values of the enthalpies of formation calculated per 1 g-atom of sulfur. The published values included in this table were taken from original papers or from textbooks [21], where these values were reviewed critically.

The enthalpies of formation of Ce_2S_3 and CeS reported in [19] were calculated using the heat of reaction between cerium and a solution of HCl given in [33]. According to a later investigation [34], this heat of reaction was too high.

TABLE 1. Enthalpies of Reactions (5), (2), and (3)

Reaction enthalpies, kJ/mole	Ln				
	La	Ce	Pr	Nd	Gd
$-\Delta H_5$	699.2±2.2	692.0±2.2	685.3±2.3	684.9±2.3	657 7±2.5
$-\Delta H_2$	247.7±2.5	243.9±2.5	240.2±2.6	241.1±2.6	237.6±2.6
$-\Delta H_3$	138.5±2.7	142.3±2.6	145.3±2.6	152.7±2.5	—

TABLE 2. Enthalpies of Formation of Rare-Earth Mono-
and Sesquisulfides

| Sulfide | Experimental results | | | Published values | |
| | $-\Delta H^0_{298}$ | | | $-\Delta H^0_{298}$ | |
	kJ g-atom	kJ mole	kcal mole	kcal mole	reference
La_2S_3	393.9	1180 ± 10	282 ± 3	282 ± 10	[21] {16, [36,37]}
Ce_2S_3	386.6	1160 ± 10	277 ± 3	300.5	[19]
				284.4	Our calc. {19,[34,35]}
Pr_2S_3	380.1	1140 ± 10	273 ± 3	—	—
Nd_2S_3	374.9	1120 ± 10	269 ± 3	285.9	[17]
				265	[21] {17, [34, 35]}
LaS	471.6	472 ± 5	113 ± 1		
CeS	468.2	468 ± 5	112 ± 1	118	[19]
				110.6	Our calc. {19,[34,35]}
PrS	465.2	465 ± 5	111 ± 1		
NdS	463.9	464 ± 5	111 ± 1		
GdS	440.2	440 ± 5	105 ± 1		

Estimates of the enthalpies of formation of Ce_2S_3 and CeS based on the results given in [19] and on the more accurate data reported in [34, 35] yielded the following values

$$\Delta H^0_{298} (Ce_2S_3) = -284.4 \text{ kcal/mole}$$

$$\Delta H^0_{298} (CeS) = -110.6 \text{ kcal/mole}.$$

These values were close to the enthalpies of formation of Ce_2S_3 and CeS obtained in our investigation.

The enthalpies of formation of La_2S_3 and Nd_2S_3 given in [21] agreed, within the limits of experimental error, with our values. According to the estimates made in [14], the covalent component of the bonding increases along the sesquisulfide series from La_2S_3 to Lu_2S_3. A similar increase in the covalent component can be expected for rare-earth monosulfides because of a reduction in the difference between the electronegativities of rare-earth metals and sulfur.

This increase in the covalence of the bonds should reduce the enthalpy of formation of rare-earth sulfides.

Our experiments (Table 2) show that the enthalpy of formation does indeed decrease with ascending atomic number of the rare-earth element. The variation of ΔH^0_{298} along the rare-earth sulfide series is correlated with the values of the melting points of LnS and Ln_2S_3 and of the activation energy of carriers in the intrinsic conduction region of Ln_2S_3.

Carter [12] suggested that the bonding in chalcogenides with the Th_3P_4 structure is predominantly of the covalent type and the bonds consist of hybrid orbitals with the C_{2v} symmetry of the electron density distribution. This type of orbital has two lobes which form half-order bonds between two different atoms. In the Th_3P_4 structure, an atom in the P position is surrounded by six Th atoms located at the vertices of a trigonal prism and it forms six bonds with these atoms by means of three C orbitals (Fig. 2). An atom of Th is surrounded by eight P atoms, to whom it is bonded by four equivalent orbitals whose lobes form two interpenetrating tetrahedra.

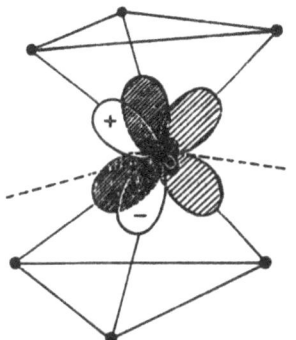

Fig. 2. Relative positions of three C orbitals forming bonds between atoms in a trigonal prism.

The covalent bonding suggests that the compounds described by Ln_2X_3 should be insulators. One-ninth of the sites occupied by the rare-earth atoms in Ln_2X_3 is empty. This gives a charge of + 2/3 to the chalcogen. The formula can therefore be written as $Ln_2^{-1} X^{+2/3}_3$ and, therefore, the balance of charge is in agreement with the observation that Ln_2X_3 compounds are insulators at 300°K. The difference between the electronegativities of the rare-earth metal and the chalcogen results in polarization and a shift of the electrons from Ln to X.

Semiconducting properties can be expected for Ln_2X_3 compounds if we start from the ionic bond assumption [38]. The formula for these compounds should then be written in the form $Ln_2^{3+} X_3^{2-}$. However, this approach ignores the directional nature of the chemical bonds.

The ionic bond approach provides a simple explanation of the metallic conduction of the trivalent (in the ground state) rare-earth monosulfides [5].

However, the covalent bond assumption seems to be more convincing. The octahedral distributions of the bonds are possible only for the d^2sp^3 and d^2sf^3 configurations [39, 40]. The bonding in the LnS-type sulfides can be represented as consisting of hybrid d^2sp^3 orbitals of the rare-earth and sulfur atoms.

Literature Cited

1. J. E. Miller, F. J. Reid, and R. C. Himes, J. Electrochem. Soc., 106:1043 (1959).
2. V. I. Kalitin, E. I. Yarembash, and N. P. Luzhnaya, Izv. Akad. Nauk SSSR, Neorg. Mater., 1:53 (1965).
3. V. P. Zhuze, A. V. Golubkov, E. V. Goncharova, T. I. Komarova, and V. M. Sergeeva, Fiz. Tverd. Tela, 6:268 (1964).
4. R. Didchenko and F. P. Gortsema, J. Phys. Chem. Solids, 24:863 (1963).
5. V. P. Zhuze, A. V. Golubkov, E. V. Goncharova, and V. M. Sergeeva, Fiz. Tverd. Tela, 6: 257 (1964).
6. R. M. Bozorth, F. Holtzberg, and S. Methiessel, Phys. Rev. Lett., 14:952 (1965).
7. V. P. Zhuze, S. S. Shalyt, V. A. Noskin, and V. M. Sergeeva, ZhETF Pis. Red., 3:217 (1966).
8. E. D. Eastman, L. Brewer, L.R.A. Bromley, P. Gilles, and N. L. Lofgren, J. Amer. Chem. Soc., 72:2248 (1950).
9. E. D. Cater, T. E. Lee, E. W. Johnson, E. G. Rauh, and H. A. Eick, J. Phys. Chem., 69: 2684 (1965).
10. W. H. Zachariasen, Acta Crystallogr., 2:57 (1949).
11. F. Holtzberg, T. R. McGuire, S. Methfessel, and J. C. Suits, Phys. Rev. Lett., Vol. 13 (1964).
12. F. L. Carter, in: Proc. Fourth Conf. on Rare Earth Research, Phoenix, Arizona, 1964 (ed. by LeRoy Eyring), Gordon and Breach, New York (1965), p. 495.
13. V. A. Obolonchik and G. V. Lashkarev, Selenides and Tellurides of Rare-Earth Metals and Actinides [in Russian], Naukova Dumka, Kiev (1966).

14. K. S. Vorres, in: Rare Earth Research II (Proc. Third Conf., Clearwater, Florida, 1963), Gordon and Breach, New York (1964), p. 147.
15. W. Blitz, Ber. Deut. Chem. Ges., 14:3341 (1908).
16. W. Blitz, Z. Anorg. Chem., 71:427 (1911).
17. C. Matignon, Ann. Chim. Phys., 8:104 (1907).
18. M. Picon and M. Patrie, C. R. Acad. Sci., p. 1321 (1956).
19. M. W. Evans, Nat. Nuclear Energy Ser., Div. IV, No. 198, p. 312 (1950).
20. H. Wartenberg, Z. Anorg. Chem., 252:136 (1943).
21. O. Kubaschewski and E. Ll. Evans, Metallurgical Thermochemistry, 3rd ed., Pergamon Press, London (1958).
22. H. Zeumer and W. Roth, Z. Elektrochem. Angew. Phys. Chem., 40:777 (1934).
23. F. D. Rossini, D. G. Wagman, W. H. Evans, S. Levine, and I. Iaffe (eds.), Selected Values of Chemical Thermodynamic Properties, Nat. Bur. Stand. (U.S.), Circ. No. 500 (1952).
24. M. M. Popov, Thermometry and Calorimetry [in Russian], Moscow State University (1954).
25. S. M. Skuratov, V. P. Kolesov, and A. F. Vorob'ev, Thermochemistry [in Russian], Vol. 1, Moscow State University (1964).
26. S. M. Skuratov, V. P. Kolesov, and A. F. Vorob'ev, Thermochemistry [in Russian], Vol. 2, Moscow State University (1966).
27. A. D. Finogenov, Determination of the Average Specific Heats of Liquids in the 20−30°C Range [in Russian], GOSINTI, Moscow (1966).
28. F. M. Jager, Ann. Phys. (Leipzig), 64:305 (1921).
29. Yu. V. Karyakin, Pure Chemical Reagents [in Russian], Goskhimizdat, Moscow (1947).
30. P. P. Korostelev, Preparation of Solutions for Chemical Analysis [in Russian], Nauka, Moscow (1966), p. 157.
31. A. V. Golubkov, T. B. Zhukova, and V. M. Sergeeva, Izv. Akad. Nauk SSSR, Neorg. Mater., 2:77 (1966).
32. S. V. Radzikovskaya and V. I. Marchenko, Sulfides of Rare-Earth Metals and Actinides [in Russian], Naukova Dumka, Kiev (1966).
33. H. Bommer and E. Hohmann, Z. Anorg. Allg. Chem., 248:357 (1941).
34. F. H. Spedding and C. F. Miller, J. Amer. Chem. Soc., 74:4195 (1952).
35. F. H. Spedding and C. F. Miller, J. Amer. Chem. Soc., 74:3158 (1952).
36. F. H. Spedding and J. P. Flynn, J. Amer. Chem. Soc., 76:1474 (1954).
37. F. H. Spedding and J. P. Flynn, J. Amer. Chem. Soc., 76:1477 (1954).
38. C. H. L. Goodman, J. Phys. Chem. Solids, 6:305 (1958).
39. G. E. Kimball, J. Chem. Phys., 8:188 (1940).
40. J. C. Eisenstein, J. Chem. Phys., 25:142 (1956).

THERMODYNAMIC PROPERTIES OF CoTe$_2$ *

V. A. Geiderikh, Ya. I. Gerasimov, and O. B. Matlasevich

Measurements were made of the emf of the cell Co(s)| CoCl$_2$ (5%), KCl−LiCl(l)| CoTe$_2$ +Te(s) in the temperature range 603-723°K. The results could be represented by the equation E = 428.9−68.4 · 10^{-3} T mV. The reaction Co(s) + 2Te(s) = CoTe$_2$(s) at 700°K was characterized by the thermodynamic parameters ΔG = −17.57 ± 0.05 kcal/mole, ΔH = −19.8 ± 0.8 kcal/mole, and ΔS = −3.2 ± 1.1 cal · deg^{-1} · mole^{-1}. The experimental values of the entropy and the enthalpy of atomization of CoTe$_2$ were analyzed on the basis of a theory of the entropy of atomization of isostructural phases and of a regular change in the enthalpy of atomization of compounds of the iron-group transition metals.

The emf method was used to investigate the thermodynamic properties of CoTe$_2$. The experimental technique was described earlier [1]. The phase diagram of the Co−Te system is shown in Fig. 1.

Measurements were made of the emf of the cell

$$(-) \ \text{Co(s)} | \ \text{CoCl}_2 \ (5\%); \ \text{KCl−LiCl}(l) | \ \text{CoTe}_2 + \text{Te(s)} \ (+) \tag{1}$$

at temperatures of 603-723°K. The results of three experiments on samples of two different compositions (15.6 and 27.6 at.% CO) are presented in Table 1.

The samples investigated were prepared by fusing together 99.9% pure tellurium and 99.2% electrolytic cobalt of the PK-1 grade. The synthesis was carried out in evacuated quartz ampoules, in which the temperature was maintained at 950°C for 3 h. This was followed by annealing for 100 h at 420°C.

The results presented in Table 1 were analyzed by the least-squares method to give the coefficients a and b in the linear temperature dependence of the emf of the cell (1): E = a + bT. In this way, we obtain the equation

$$E = 428.9 − 68.4 \cdot 10^{-3} T. \tag{2}$$

The variances (s^2) were determined for each measurement of E(s$_E^2$), for the coefficients a(s$_a^2$) and b(s$_b^2$), and for the value of $\tilde{E}$ calculated from Eq. (2) (s$_{\tilde{E}}^2$). These variances were found by regression analysis (see, for example, [3, 4]). The error was found by doubling the absolute value of the square root of the variance (confidence interval 95%): ±2s$_E$ = ±3.5 mV; ±2s$_a$ = ±16.4 mV; ±2s$_b$ = ±0.024 mV/deg; ±2s$_{\tilde{E}}$ = ±2|[0.1255 + 1.48 · 10^{-4} (T − 671.9)^2]$^{1/2}$|.

The potential recorded in the cell (1) resulted from the formation of CoTe$_2$ from its com-

---

* "Crystals," pp. 292-295 (see page 3).

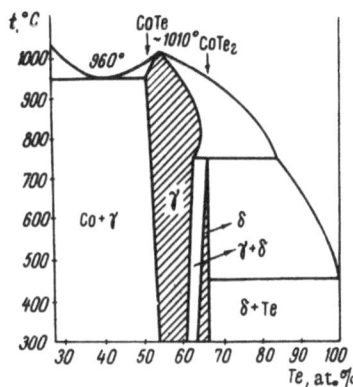

Fig. 1. Phase diagram of the
Co—Te system [2].

ponents in the solid state

$$Co \text{ (s)} + 2Te \text{ (s)} = CoTe_2 \text{ (s)}. \qquad (3)$$

The thermodynamic functions of the reaction (3) were related to the emf of the cell (1) by the well-known equations

$$\Delta G = -zFE, \qquad (4)$$

$$\Delta S = zF\left(\frac{\partial E}{\partial T}\right)_p, \qquad (5)$$

$$\Delta H = zF\left[T\left(\frac{\partial E}{\partial T}\right)_p - E\right], \qquad (6)$$

where the Faraday is $F = 23.061$ cal/mV and the charge of the ion responsible for the measured potential is $z = 2$.

The following values were obtained for the thermodynamic functions of the reaction (3) at 700°K: $\Delta G = -17.57 \pm 0.05$ kcal/mole; $\Delta H = -19.8 \pm 0.8$ kcal/mole; $\Delta S = -3.2 \pm 1$ cal·deg^{-1}·mole^{-1}. When these functions were recalculated per 1 g-atom (1/3 $CoTe_2$), we obtained: $\Delta G = -5.86 \pm 0.02$ kcal/g-atom; $\Delta H = -6.6 \pm 0.3$ kcal/g-atom; $\Delta S = -1.1 \pm 0.4$ cal·deg^{-1}·g-atom^{-1}.

No published data were available on the specific heat of $CoTe_2$. It was known that $C_{p298} = 17.60$ cal·deg^{-1}·mole^{-1} for $FeTe_2$ [5]. We used the data in [6] on the specific heats of the components and these gave us $\Delta C_{p298}(FeTe_2) = -0.75$ cal·deg^{-1}·mole^{-1}. We assumed that ΔC_p of the compounds $FeTe_2$ and $CoTe_2$ did not differ very greatly and, therefore, that $\Delta C_p(CoTe_2) \approx -0.75$ cal·deg^{-1}·mole^{-1}. Assuming that $d(\Delta C_p)/dT = 0$, we obtained

$$\Delta H_{298} = \Delta H_T - \Delta C_p(T - 298), \qquad (7)$$

$$\Delta S_{298} = \Delta S_T - \Delta C_p \cdot 2.3 (\log T - \log 298). \qquad (8)$$

The average temperature for the investigated range was $T = 672$°K. We referred the experimental data to this temperature and substituted the numerical values in Eqs. (7) and (8). This gave

$$\Delta H_{298}(CoTe_2) = -19.5 \text{ kcal/mole} = -6.5 \text{ kcal/g-atom}$$

$$\Delta S_{298}(CoTe_2) = -2.5 \text{ cal·deg}^{-1}\text{·mole}^{-1} = -0.8 \text{ cal·deg}^{-1}\text{·g-atom}^{-1}$$

The errors in such calculations were probably not greater than the errors in the experimental values of ΔH and ΔS. The enthalpy of the formation of $CoTe_2$ from its components

TABLE 1. Temperature Dependence of emf's

at.% Co	T, °K	E, mV	at.% Co	T, °K	E, mV
15.6	603	389.19	15.6	671	384.80
27.6	634	384.60	27.6	671	383.90
27.6	639	383.53	27.6	681	383.75
15.6	639	383.08	15.6	681	384.06
15.6	639	385.01	15.6	684	379.23
27.6	655	384.91	27.6	688	383.19
27.6	660	384.06	15.6	688	383.99
15.6	667	379.29	15.6	706	380.00
15.6	668	382.99	27.6	707	381.10
27.6	668	382.57	15.6	707	381.33
15.6	668	384.37	15.6	711	381.10
27.6	668	384.31	15.6	723	376.23

(ΔH_{298}) differed considerably from the value -31.5 ± 0.8 kcal/mole, which was obtained by Ariya et al. [7] from the difference between the enthalpies of the interaction of cobalt ditelluride with hydrochloric acid and bromine and the corresponding mixture of the components. The reason for this large discrepancy was not clear.

It seemed of interest to compare the thermodynamic properties of $CoTe_2$ with the properties of similar compounds in the systems investigated earlier (Fe–Te, Cr–Sb, Fe–Sb, and Co–Sb). The most convenient approach was to compare the entropies and enthalpies of atomization whose values were calculated for $CoTe_2$ from the experimental results using the following data for the components taken from [6]: ΔH_{298}^{at} $(CoTe_2) = 71.4$ kcal/g-atom; $\Delta S_{298}^{at}(CoTe_2) = 33.9$ cal $\cdot$ deg^{-1} $\cdot$ g-atom^{-1}.

Let us consider the values obtained on the basis of the hypotheses which we made in [8, 9].

In [8], we analyzed the entropies of atomization of phases with structures of the following types: NiAs, marcasite, pyrite, and NaCl. We advanced the hypothesis that the entropies of atomization of isostructural phases were similar in those cases when the phases contained a considerable contribution from the metallic (cooperative) interaction. In particular, the phases with the marcasite-type structure ($CoTe_2$ also has this type of structure) had values of ΔS^{at} which lay close to the average value $\Delta S_{298}^{at} = 34.4$ cal $\cdot$ deg^{-1} $\cdot$ g-atom^{-1} (the scatter of these values was ± 1.1 cal $\cdot$ deg^{-1} $\cdot$ g-atom^{-1}). The value $\Delta S_{298}^{at} = 33.9$ cal $\cdot$ deg^{-1} $\cdot$ g-atom^{-1} obtained in the present investigation for $CoTe_2$ differend by only 0.5 cal $\cdot$ deg^{-1} $\cdot$ g-atom^{-1} from ΔS_{298}^{at} of marcasite, i.e., it was in good agreement with the hypothesis put forward in [8].

In [9], we compared the values of ΔH^{at} of the compounds of transition metals with Al, Sb, and Sn, and we found that the enthalpy of atomization of these compounds increased along the iron–cobalt–nickel series. This was compared with the postulated rise of the electron affinity along the same series of the iron-group transition elements. In [10], we drew attention to the fact that the same relationship was obeyed by silicon alloys rich in transition metals (these alloys were characterized by relatively strong metallic interaction). This relationship was not obeyed by the compounds of transition metals with nonmetals (such as transition-metal sulfides).

Iron, cobalt, and nickel ditellurides are known to have a mixed type of binding with a considerable contribution from the covalent interaction (see, for example, [11]). The ionization potential of tellurium (8.96 eV [13]) is higher than that of antimony and only 1.3 eV lower than that of sulfur.

In view of this, it would be interesting to find whether the enthalpy of atomization of ditellurides increases along the iron–cobalt–nickel series. There are no published data on the enthalpy of formation of $NiTe_2$. The value of ΔH_{298}^{at} of $FeTe_2$ is 69.6 kcal/g-atom (this value was calculated by us from ΔH of $FeTe_2$ [12] and from ΔH^{at} of the components [6]). Thus, we can only say that ΔH_{298}^{at} of $CoTe_2$ is greater than ΔH_{298}^{at} of $FeTe_2$.

Literature Cited

1. Ya. I. Gerasimov, A. V. Nikol'skaya, V. A. Geiderikh, A. S. Abbasov, and R. A. Vecher, in: Chemical Bonds in Semiconductors and Solids (ed. by N. N. Sirota), Consultants Bureau, New York (1967), p. 87.
2. L. D. Dudkin and K. A. Dyul'dina, Zh. Neorg. Khim., 4:2313 (1959).
3. A. N. Kornilov, Zh. Fiz. Khim., 41:3096 (1967).
4. V. V. Nalimov, Applications of Mathematical Statistics in Analysis of Matter [in Russian], Moscow (1960).
5. E. F. Westrum, Jr., Chien Chou, and F. Grønvold, J. Chem. Phys., 30:761 (1959).

6. D. R. Stull and G. C. Sinke, Thermodynamic Properties of the Elements, American Chemical Society, Washington (1956).
7. S. M. Ariya, E. M. Kolbina, and M. S. Apurina, Zh. Neorg. Khim., 2:23 (1957).
8. V. A. Geiderikh and Ya. I. Gerasimov, Zh. Fiz. Khim., 41:2601 (1967).
9. V. A. Geiderikh and Ya. I. Gerasimov, Zh. Fiz. Khim., 37:2353 (1963).
10. R. A. Vecher, V. A. Geiderikh, and Ya. I. Gerasimov, Izv. Akad. Nauk SSSR, Neorg. Mater., 1:1722 (1965).
11. T. Rosenqvist, Magnetic and Crystallographic Studies of the Higher Antimonides of Iron, Cobalt, and Nickel, Trondheim (1953).
12. V. A. Geiderikh, Ya. I. Gerasimov, and A. V. Nikol'skaya, Dokl. Akad. Nauk SSSR, 137:1399 (1961).
13. Handbook of Chemistry and Physics, 37th ed., The Chemical Rubber Co., Cleveland, Ohio (1955-6).

THERMODYNAMIC PROPERTIES OF CRYSTALS IN RELATION TO THE NATURE AND ENERGY OF THE INTERATOMIC INTERACTION*

N. N. Sirota

The effect was investigated of the nature and energy of the interatomic interaction on the structure and physical properties of crystals. Factors were studied which determine the thermodynamic properties and their temperature dependences. The effect of various parameters on the form of the frequency spectrum of phonons was investigated, taking, as an example, crystals with a diamond-type structure. The problem was also studied of the determination of the elastic constants as derivatives of the crystal energy in terms of the distribution functions of the electron density, represented by various approximations.

One of the main problems in crystal chemistry, crystal physics, and crystallography in general is the question of the chemical bonding in crystals and the effect of the nature and energy of the interatomic interaction on their structure and physical properties. Many physical properties of crystals are described by the appropriate derivatives of their thermodynamic properties. For this reason, a study of especial interest is that concerning the factors determining the thermodynamic properties of crystals and their dependence on generalized forces and coordinates. These thermodynamic properties are, in particular, the atomization energy, the internal energy, the Helmholtz free energy, the Gibbs free energy, the specific heat, and others.

If the zero-point atomization energy, the heats of formation and their dependences on temperature and on the generalized forces and coordinates are known, all the most important thermodynamic functions and their temperature dependences can be determined.

Later on, we shall consider some of the main factors determining the lattice components of the thermodynamic properties of monatomic crystals.

First of all, we shall record the experimental facts connecting the crystal structure with the temperature dependence of the specific heat.

The specific heat of crystals is dependent on the vibration specific heat of the ions, the electron specific heat, the specific heat due to diffusion of atoms, the specific heat due to defect formation, etc. In particular, the electron specific heat is proportional to the temperature; the specific heat due to diffusion and self-diffusion is an exponential function of the temperature. For example, if the number of ion movements in the crystal lattice in a volume of one gram-atom in unit time is $N_A = \nu N e^{-U/kT}$, and in each migration energy is expended in overcoming

---

* "Semiconductors," pp. 183-194 (see page 3).

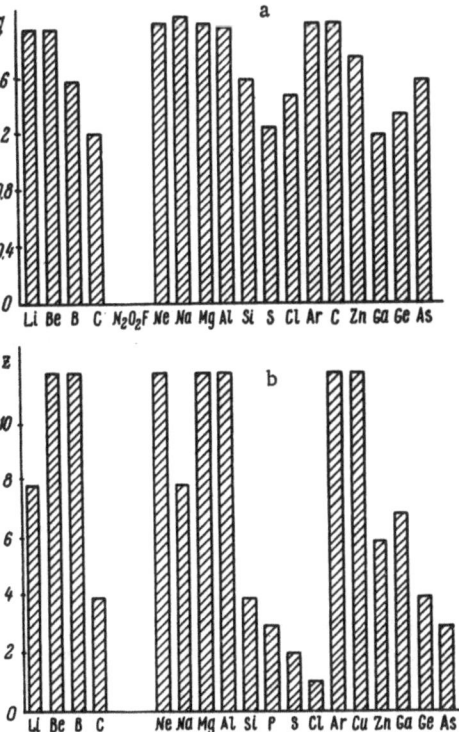

Fig. 1. Dependence of the exponent n = q + 1 in the temperature dependence of the specific heat on the position of an element in the periodic table (a) and dependence of the effective coordination number z in the unit cell of an element (b).

the energy threshold U or in unit time $\varepsilon = N\nu U e^{-U/kT}$, then in this case the specific heat due to diffusion is

$$c_D = \frac{d\varepsilon}{dT} = \frac{NeU^2}{kT^2}\, e^{-\frac{U}{kT}}.$$

Let us next consider the lattice specific heat.

The specific heat of a large number of solids is fairly well described by the Debye interpolation equation $c_V = f(T/\Theta)^3$, according to which, particularly at low temperature, the familiar Debye cube law $c_V = a(T/\Theta)^3$ is obeyed.

However, Lewis and Gibson [1], Eucken [2], and others showed that there are a large number of substances including elements, for which the temperature dependence of the specific heat is not described by the Debye function over a wide temperature range. For example, as can be seen from Fig. 1, deviations from the Debye law are particularly noticeable for elements whose p-electron shells are being occupied; the symmetries and lattice coordination numbers z of these elements depend on the extent to which the p shells are occupied.

Using the well-known approximation, the temperature dependence of the specific heat of such substances can be described by interpolation equations of the form [3] $c_V = f(T/\Theta)^{q+1} = f(T/\Theta)^n$.

Using Lewis and Gibson's method [1], the change in the exponent n is shown in Fig. 1 for elements of the second and third periods. As is seen from the figure, there is a regular variation in n depending on the group number of the element, passing through a minimum in the region where the p-electron shells are being occupied. To stress the connection with the anisotropy of the structure, the change in the effective coordination number z is also shown. By comparing the data, it can be seen that, when the coordination number falls, the exponent n tends to decrease.

The deviations from the Debye law are, undoubtedly, connected with the structure of the electron shells of atoms and their shape. By the shape of the atom or ion, one may understand the shape of the surface of equal electron density of the outer-shell electrons of the atom. From this point of view, the ions whose outer electrons are the s electrons have spherical symmetry, whereas the p electrons and d electrons form dumb-bells, the shapes of which are shown in Fig. 2a according to Hume-Rothery. The shapes of the p and d shells cannot, obviously, be approximated by spheres.

The lattices of the elements whose p-electron shell is being occupied generally obey the Hume-Rothery law z = 8 − N (z is the coordination number and N is the group number in the periodic table).

A change in the nature of the bonds occurs on passing from the elements whose s-electron shells are being occupied to the elements whose p-electron shells are being occupied. For these elements, p-p, s-s, and s-p directional bonds are the characteristic bonds; the form of these bonds is shown in Fig. 2b for diamond and selenium.

As has been shown previously by us [3] and independently by Tarasov [4], an allowance for the regularity of elastic wave propagation in crystals gives a Debye-type expression in which the exponent is one (n = 1) for one-dimensional structures (filaments), two (n = 2) for laminar structures, and three (n = 3) for isotropic space structures. In all the intermediate cases, the exponent n will have an intermediate value. However, in contrast to Tarasov, we did not consider the behavior of filaments or layers and their interaction with each other, but the special features of occupation by figurative points corresponding to the excitation of vibrations in the phase k space of anisotropic substances.

The effect of the nature of the covalent, ionic, or van der Waals bond is clearly apparent in the deviations of the temperature dependences of the specific heat from the Debye law. The bond type causes a change in the anisotropy mainly in the vibration component of specific heat. The directional nature of covalent bonds leads to a reduction in the exponent n.

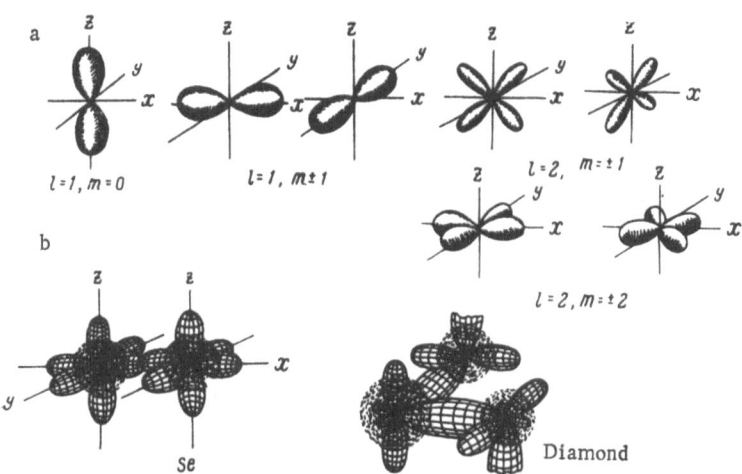

Fig. 2. Three-dimensional orientation of electron clouds.

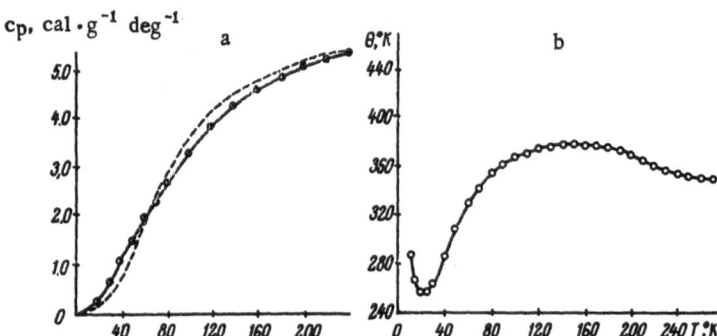

Fig. 3. Temperature dependences of specific heat (a)
and characteristic temperature (b) of germanium (the
dashed curve represents the specific heat according
to the Debye law).

For elements with s, d, and f shells forming metallic bonds there is an electron component of the specific heat, including that determined by transitions in d and f shells, with its specific temperature dependence.

The above-mentioned relationships may be due to the form of the initial part of the phonon spectrum, i.e., its acoustical branch: for three-dimensional Debye solids the spectrum is $g_3(\nu) \propto k_3\nu^2$, for two-dimensional structures $g_2(\nu) \propto k_2\nu$, and for one-dimensional structures $g_1(\nu) \propto k_1$.

Of course, these relationships and the Debye equation itself are only approximate. However, the merits of the Debye method cannot be ignored. The distinguishing feature and, partly its advantage is that it considers a solid as a continuum; it determines the natural frequencies of this continuum and thereby formulates a definite statistical interpretation of the isotropic filling of the k phase space independent of the actual structure of the crystal. In addition, this method makes it possible to consider approximately also the anisotropy of the filling of the k space.

In addition, the connection between the characteristic temperature and the elastic constants and the heat of atomization can be established directly by the Debye method. We note, for example, that the Debye characteristic temperature is directly proportional to the average velocity of elastic waves in a crystal ($\Theta = \overline{kc}$) and it is proportional to the square root of the atomization energy per unit atomic weight, at least in the first approximation ($\Theta = k\sqrt{U/A}$). The Debye characteristic temperature of isotropic bodies is directly associated with the bulk modulus and therefore with all the other properties of matter dependent, to a certain degree, on the atomization energy ΔH_{at}, the surface energy σ, the elasticity moduli, the expansion coefficients, the width of the forbidden band in semiconductors, etc.

As we have previously shown [5], there is also a definite relationship between the Debye characteristic temperature and the heat of activation of diffusion U, etc. However, an analysis of the experimental curves of the specific heat as a function of temperature shows that, for all bodies, including those whose specific heat is more or less satisfactorily described by the Debye equation, the characteristic temperature of the whole crystal is essentially a function of temperature (Fig. 3). In addition, it is observed that the greater the deviation of the curve of Θ from the horizontal straight line, the more the temperature dependence of the specific heat deviates from the Debye law. Despite the fact that the temperature dependence of the specific heat is comparatively insensitive to the form of the phonon spectrum, an evaluation of the trend of the specific heat curves already indicates that the vibration spectrum of ion vibrations in real solids differs essentially from the Debye law.

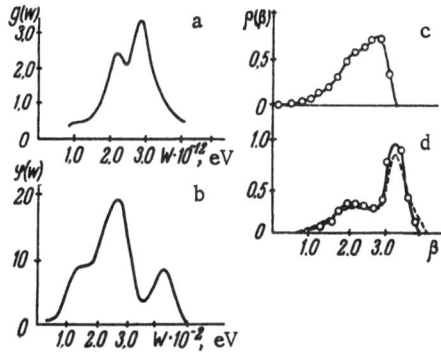

Fig. 4. Phonon spectra of W (a), Mn–Co (b), Be (c), and BeO (d), determined experimentally by inelastic neutron scattering [6, 8–10, 12].

Experimental investigations by Brockhouse and others [6–11] on cold-neutron scattering by the time–of–flight method or by diffraction methods, and also work by Cribier and others [12–14] using x-ray diffraction, have made it possible to determine almost exactly the true form of the phonon spectra.

Some of the phonon spectra, determined experimentally, are shown in Fig. 4.

A knowledge of the vibration spectrum of oscillation of ions and electrons in a crystal at different temperatures and external fields solves, to some extent, the problem of the thermodynamic properties of solids. The value of the internal energy U, the specific heat c_V, and other thermodynamic functions is expressed directly in terms of the density of the frequency distribution $g(\nu)$ and the average energy of the oscillators:

$$U = U_0 + \int_0^\infty g(\nu)\,\bar{\varepsilon}\,d\nu, \quad c_V = \frac{\partial}{\partial T} \int_0^\infty g(\nu)\,\bar{\varepsilon}\,d\nu, \quad \ldots,$$

$$F = U_0 - \int \frac{dT}{T^2} \int c_V\,dT, \text{ etc.}$$

The average energy of the oscillator $\bar{\varepsilon}$, according to Planck, is

$$\bar{\varepsilon} = \frac{h\nu}{e^{\frac{h\nu}{kT}} - 1} + \frac{h\nu}{2}.$$

Let us consider in more detail the peculiarities of the vibration spectra of crystals, limiting the investigation to ion oscillations, mainly in cubic crystals.

We will consider the change in the phonon spectrum of a crystal with a diamond-type structure, whose ionic mass (isotopic effect) and lattice constant change, but whose elastic constants remain the same. We will give the calculated phonon spectra of a crystal with a diamond-type structure, with different atomic masses. The calculation scheme was similar to that in [15].

A change in the ion mass and the lattice constants causes only a change in the scale along the frequency axis (Fig. 5). If the limiting frequency is taken as unity, the external form of the phonon spectrum is unchanged by variation of the ion mass and the lattice constant.

However, the temperature dependences of the specific heat, internal energy, free energy, and other thermodynamic properties do, in this case, change; an increase in the ion mass is inversely proportional, and an increase in the lattice constant is directly proportional to the square of the limiting frequency, which must be considered as an analog of the characteristic temperature.

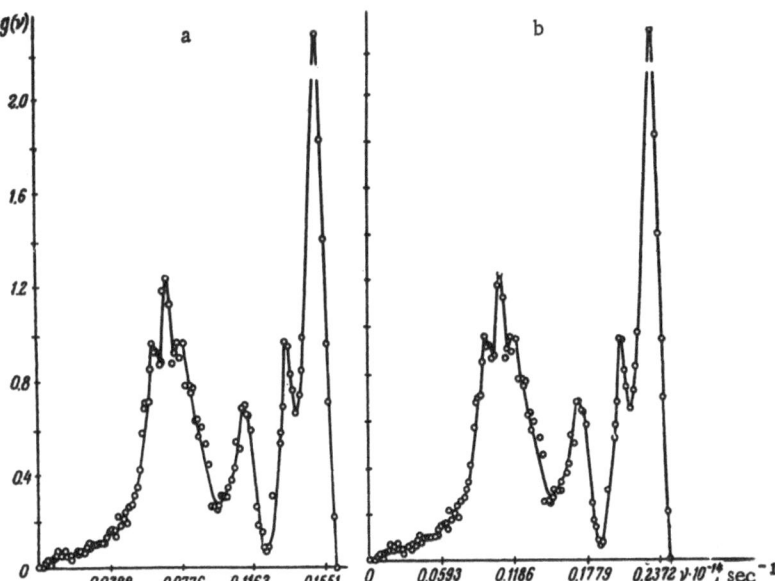

Fig. 5. Phonon spectra of crystals with a diamond-type struc-
ture, plotted for $m_1 = m_{Si}$ (a) and $m_2 = 0.43 \, m_{Si}$ (b); $g(\nu)$ is in
relative units.

Thus, provided the elastic properties remain constant, a change in the lattice constant and
the ion mass does not change the temperature dependence of the specific heat. If graphs are
plotted of T/Θ for different values of m and a, they will be superimposed on each other.

Let us consider the effect of the degree of elastic anisotropy of the lattice of, for example,
a diamond-type crystal on the form of the frequency spectrum and on the temperature dependence
of the specific heat and other thermodynamic functions. In order to see the effect of such an
anisotropy, we will vary the anisotropy coefficient $A = 2c_{44}/(c_{11} - c_{12}) = 2y$ when the ratios of
the bulk moduli to the shear moduli are constant:

$$\frac{(c_{11} + 2c_{12})_I \, k_I}{(c_{11} + 2c_{12})_{II} k_{II}} = \beta = \text{const} \quad \text{and} \quad \frac{(c_{44})_I}{(c_{44})_{II}} = \text{const.}$$

The phonon spectra of the diamond lattice obtained for a single limiting frequency, dif-
ferent values of the anisotropy constant, and $k_I/k_{II} = \text{const}$, $(c_{44})_I/(c_{44})_{II} = \text{const}$, are characterized by the fact that an increase in the coefficient $y = c_{44}/(c_{11} - c_{12})$ alters significantly the
density of the frequency distribution in the low-frequency part of the acoustical branch, changing it from an almost linear dependence at $y = 0.2$ to a parabolic dependence at $y = 0.9$ (Fig. 6).

An analysis of the frequency dependence of the distribution of the lowest frequencies over
a range of less than one tenth of the limiting frequency showed that in many of the investigated
cases a section of the curve may be found where the dependence $g(\nu) \propto \nu^2$ is obeyed.

Some interesting and important conclusions were drawn by separating the phonon spectrum
in accordance with the polarization of the oscillations [15]. The whole spectrum was divided into
six branches, each of which has an almost Gaussian form of the distribution curve $g(\nu)$. For
cubic crystals, these six branches consist of three acoustical branches (one branch of longitudinal and two branches of transverse waves) and three optical branches (one longitudinal and two
transverse waves). The acoustical vibrations can be compared with the vibrations of atoms in
a unit cell, and the optical vibrations with mutual oscillations of the sublattices in relation to
one another. The curves of the density distribution of oscillations in each of the enumerated

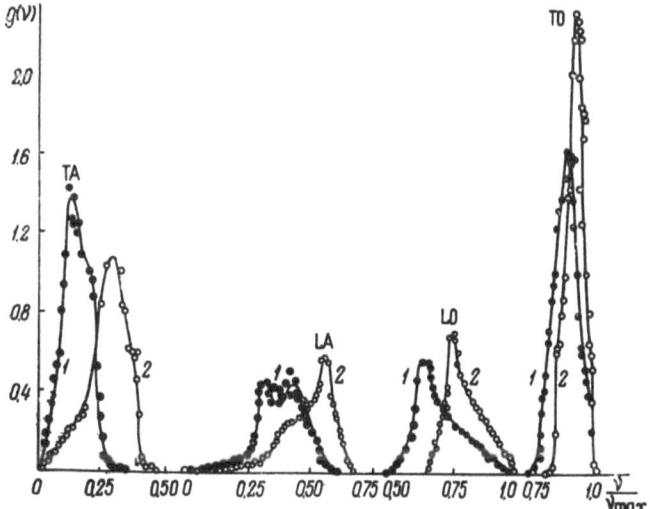

Fig. 6. Distributions of transverse and longitudinal acoustical (TA, LA) oscillations and longitudinal and transverse optical (LO, TO) oscillations for y = 0.2 (1) and 0.8 (2); β = 1 in both cases.

branches are Maxwellian or Gaussian (Fig. 6). For each of these branches the maximum of the distribution density is characterized by a specific frequency. Since, in cubic crystals, two acoustical branches of transverse vibrations comprise pairs which almost coincide, the temperature dependence of the specific heat is fully described by four characteristic temperatures. However, the temperature dependence of the specific heat is mainly determined by the acoustical vibrations; therefore, in this case, it is in practice sufficient to know two characteristic temperatures which are independent of temperature.

From the trend of the temperature dependences of the specific heat, plotted on a reduced scale (Fig. 7) in which a temperature corresponding to a specific heat of 3 cal · mole^{-1} · deg^{-1} is taken as unity, it can be seen that when the coefficient $y = c_{44}/(c_{11} - c_{12})$ is increased the curves approach the Debye function. Similar behavior is exhibited by the temperature dependences of the Helmholtz free energy and the entropy for different values of y, plotted on the same scale (Figs. 8, 9).

Our analysis of the phonon spectra and of the temperature dependences of the specific heat indicates that, proceeding from an investigation of the temperature dependence of the specific heat and the form of the frequency spectrum, definite crystal-chemistry conclusions can be drawn

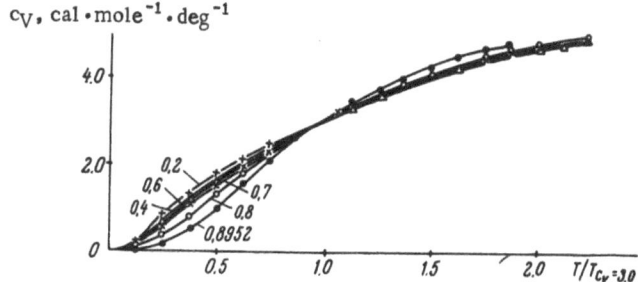

Fig. 7. Temperature dependences of the specific heat for different values of the coefficient y.

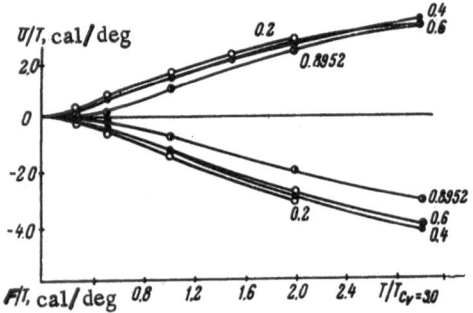

Fig. 8. Temperature dependences of the Helmholtz free energy F and the internal energy U, related to the temperature for which $c_V = 3$, for different values of the coefficient y.

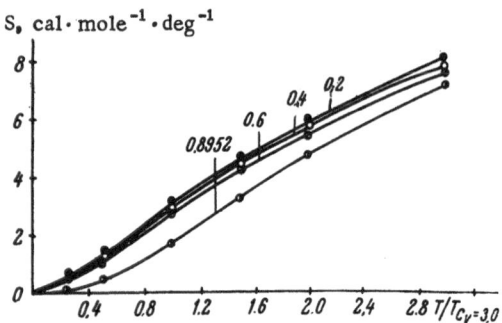

Fig. 9. Temperature dependences of the entropy S for different values of the coefficient y.

concerning the nature and energy of the atomic bonds and certain peculiarities of the crystal structure, particularly its anisotropy.

As we mentioned above, the force constants or the elastic constants are used in calculating the phonon spectrum.

For crystals with a cubic structure, the elastic properties are characterized by three elastic moduli: c_{11}, c_{12}, and c_{44}. When considering the first two coordination spheres in the central-force approximation, these three moduli are sufficient to reproduce the whole phonon spectrum. When the next coordination spheres are considered, additional force constants must be known. The number of force constants which must be known depends on the number n of the spheres being considered.

The values of the elastic moduli and their relationships are unequivocally determined by the structure and the nature and energy of the atomic chemical bonds. Since the structure and also the nature and energy of the bonds depend on the electron density distribution in the crystal, the elastic moduli also depend ultimately on its distribution.

The author would like to thank T. D. Sokolovskii and N. S. Orlova for their help in the calculations.

Literature Cited

1.   G. N. Lewis and G. E. Gibson, J. Amer. Chem. Soc., 39:2554 (1917).
2.   A. Eucken, Lehrbuch der Chemischen Physik, Akademische Verlagsgesellschaft, Leipzig (1944).
3.   N. N. Sirota, Dokl. Akad. Nauk SSSR, 47:40 (1945).
4.   V. V. Tarasov, Dokl. Akad. Nauk SSSR, 46:22 (1945).
5.   N. N. Sirota, in: Chemical Bonds in Semiconductors and Thermodynamics (ed. by N. N. Sirota), Consultants Bureau, New York (1968), p. 1.
6.   A. T. Stewart and B. N. Brockhouse, Rev. Mod. Phys., 30:250 (1958).
7.   K. C. Turberfield and P. A. Egelstaff, Phys. Rev., 127:1017 (1962).
8.   N. A. Chernoplekov, M. G. Zemlyanov, and A. G. Chicherin, Zh. Eksp. Teor. Fiz., 43: 2080 (1962); N. A. Chernoplekov, M. G. Zemlyanov, E. G. Brovman, and A. G. Chicherin, in: Inelastic Scattering of Neutrons in Solids and Liquids, Vol. 2, International Atomic Energy Agency, Vienna (1963), p. 173.

9. B. N. Brockhouse and P. K. Iyengar, Phys. Rev., 111:747 (1958).
10. C. M. Eisenhauer, I. Pelah, D. J. Hughes, and H. Palevsky, Phys. Rev., 109:1046 (1958).
11. G. Dolling and R. A. Cowley, Proc. Phys. Soc., London, 88:463 (1966).
12. D. Cribier, B. Jacrot, and D. Saint-James, J. Phys. Radium, 21:67 (1960).
13. D. Cribier, Acta Crystallogr., 6:293 (1953); Ann. Phys. (Paris), 4:333 (1959).
14. H. Curien, Bull. Soc. Fr. Miner., 75:197, 343 (1952); Acta Crystallogr., 5:393 (1952).
15. N. N. Sirota and T. D. Sokolovskii, in: Chemical Bonds in Semiconductors and Thermodynamics (ed. by N. N. Sirota), Consultants Bureau, New York (1968), p. 139.

THERMODYNAMIC PROPERTIES OF AlSb AND AlSb—GaSb SOLID SOLUTIONS*

V. V. Samokhval, A. A. Vecher, and E. P. Pan'ko

The thermodynamic properties were determined of AlSb – GaSb solid solutions, with 10 at. % excess antimony, by measuring the electromotive forces of the galvanic cells |AlSb, Sb, CaAlF₅, CaF₂|CaF₂|(Al, Ga)Sb, Sb, CaAlF₅, CaF₂| at 855°K. Negative deviations from Raoult's law were found.

Thermodynamic data are frequently used in an analysis of the laws governing the change of properties of $A^{III}B^V$ semiconducting compounds with change of position of the constituent elements in the periodic table [1–3]. To resolve this problem, it is necessary to obtain sufficiently reliable data on the thermodynamics of the formation of such compounds.

The thermodynamic properties of aluminum antimonide have been previously studied by the electromotive force (emf) method using aluminum chloride as the electrolyte in fused lithium and potassium chlorides [4]. The calculated value [4] of the standard entropy of solid aluminum antimonide $S^0_{298} = 6.0 \pm 0.8$ eu/g-atom and that obtained by Piesbergen [5] from measurements of low-temperature specific heat, $S^0_{298.16} = 7.68 \pm 0.05$ eu/g-atom do not agree even within the limits of experimental error (here, eu = entropy unit = cal/deg). For the isovalent solid solutions of $A^{III}B^V$ semiconductors, there are no published data on the free energy and the energy of their formation.

We studied the thermodynamic properties of aluminum antimonide and the AlSb – GaSb system by the emf method, using a solid electrolyte.

The emf's of the following galvanic cells were measured

$$|Al, \ CaAlF_5, \ CaF_2| \ CaF_2 \ |AlSb, \ Sb, \ CaAlF_5, \ CaF_2| \qquad (1)$$

in the 778 to 895°K temperature range, and

$$|AlSb, \ Sb, \ CaAlF_5, \ CaF_2| \ CaF_2| \ (Al, \ Ga) \ Sb,$$
$$Sb, \ CaAlF_5, \ CaF_2| \qquad (2)$$

at 855°K.

The potential-forming reactions of cells (1) and (2) are, respectively,

$$Al_{solid} + Sb_{solid} = AlSb_{solid}. \qquad (3)$$

$$AlSb_{solid} = AlSb_{in \ the \ solid \ solution} \qquad (4)$$

---

* "Semiconductors," pp. 195-199 (see page 3).

TABLE 1. Heats of Formation (kcal/mole) of Phosphides, Arsenides, and Antimonides of Aluminum, Gallium, and Indium from the Components, Taken in the Ideal Monatomic Gas State (Heats of Atomization)

Element	P	As	Sb
Al	186.85 [6]	172.35 [7]	143.05
Ga	168.80 [8]	151.7 [8]	138.4 [8]
In	157.4 [9]	139.8 [9]	116.5 [10]

N o t e . Sources are shown in the table from which data were taken on the heats of formation of compounds from their elements; heats of evaporation were taken from [11]; the heat of formation of AlSb is from our results.

From the equation of the dependence of the emf of cell (1) on temperature,

$$E = 185.9 - 49.7 \cdot 10^{-3} T \pm 1.5 \text{ mV},$$

the Gibbs energy ΔG, entropy ΔS, and enthalpy ΔH were calculated for the investigated temperature range. A calculation for 800°K gave the following results :

$$\Delta H^0_{800} = -12.86 \pm 0.19 \text{ kcal/mole}$$

$$\Delta S^0_{800} = -3.44 \pm 0.22 \text{ eu/mole,}$$

$$\Delta G^0_{800} = -10.11 \pm 0 10 \text{ kcal/mole.}$$

According to Piesbergen's data [5] for the reaction in which AlSb is formed, $\Delta C_p = -0.83$ cal · deg^{-1} · mole^{-1} at 298°K. Using our data and assuming that ΔC_p is independent of temperature, we obtain

$$\Delta H^0_{298} = -12.44 \text{ kcal/mole}$$

$$\Delta S^0_{298} = -2.60 \text{ eu/mole,}$$

$$\Delta G^0_{298} = -11.67 \text{ kcal/mole.}$$

The calculated values of the standard entropy of AlSb $S^0_{298} = 7.53 \pm 0.25$ eu/ g-atom from our data and $S^0_{298.16} = 7.68 \pm 0.05$ eu/ g-atom from Piesbergen's data agree within the limits of experimental error.

It can be seen from Table 1 that, in the homologous series of the type AlP, AlAs, AlSb (with a common group III element) or the type AlP, GaP, InP (with a common group V element), a regular decrease is observed in the heat of atomization on increasing the atomic number of one of the components. Since all these compounds have sphalerite-type lattices, weakening of the bonds is caused by spreading out of the covalent "bridges" because of greater screening of the nucleus [12].

In the series of diagonal electron analogs consisting of elements with a similar combined atomic number (for example, InP, GaAs, AlSb), the heat of atomization decreases as the atomic number of the "cation" is reduced. The GaP, AlAs series is an exception to this rule. On the basis of data for heats of atomization obtained by a simple interpolation of values for neighbors in the horizontal series, Ormont [13] came to the opposite conclusion.

Our analysis shows that thermodynamic properties obtained for aluminum antimonide agree satisfactorily with general rules for semiconductors of this group [14]. This was the reason for using the emf of galvanic cells, with a solid electrolyte, for determining the thermodynamic properties of the AlSb−GaSb system, in which solid solutions are formed by isovalent substitution [15, 16].

The emfs of cells (2) were measured at a constant temperature for seven different compositions. The standard electrode was prepared by grinding powders of aluminum, antimony, and the $CaAlF_5 - CaF_2$ salt mixture, followed by compacting. In all the experiments, there was a 20 at.% excess of antimony with respect to the stoichiometric composition, AlSb. The compacted pellet was then annealed for 10 h at 800°C in an evacuated and sealed quartz ampoule. The electrode melt was prepared by compacting the previously obtained AlSb−GaSb alloy with the salt mixture. Alloys enriched with AlSb were handled in a dry cell. Samples of the

AlSb—GaSb solid solutions were obtained by melting the appropriate amounts of aluminum
(AV000), antimony (Su000), and gallium (Gl0), there being a constant 10 at.% excess of antimony.
Melting was carried out in boron carbonitride crucibles in evacuated and sealed quartz ampoules
at temperatures 100 deg above the liquidus; the temperature was then slowly reduced below the
solidus, and the samples were annealed for 15 h. There were no traces of evaporated material
on the ampoule walls. A further anneal was carried out in the measuring cell itself.

A monocrystalline plate of calcium fluoride, 1 to 1.5 mm thick, was used as the electrolyte.
Experiments were carried out in a vacuum of better than $2 \cdot 10^{-4}$ torr. The equilibrium value
of the emf was reached after 1–3 days, depending on the composition.

The change of the partial molar free energy of AlSb was calculated from the formula

$$\Delta \bar{G}_{AlSb} = -3FE, \tag{5}$$

where F is the Faraday number; E is the emf. The AlSb activity was determined from the relationship

$$\ln a_{AlSb} = -\frac{\Delta \bar{G}_{AlSb}}{RT} \tag{6}$$

The corresponding values for GaSb were calculated from the Gibbs—Duhem equation.

The emf's of cell (2) are shown in Table 2. The emf's of four different cells were measured for each composition. The scatter in the values was 5%. This is probably explained by
a certain nonuniformity of the alloys, since even after zone leveling it is difficult to obtain
homogeneous samples in this system [16]. As can be seen in Table 2, strong interaction between the components (AlSb and GaSb) is observed in the AlSb—GaSb system.

For the Ge—Si, InSb—GaSb, InSb—AlSb, InSb—InAs, InAs—GaAs, and InAs—InP systems, a
theoretical calculation has been made [17, 18] of constitutional diagrams on the regular-solution
approximation. The value and sign of the displacement energy used by these authors [17, 18]
as giving the best agreement between calculated and experimental constitutional diagrams indicate a tendency toward stratification in these systems, although this has not been detected
experimentally.

On the basis of our measurements, we can now say with certainty that AlSb—GaSb solid
solutions with excess antimony deviate negatively from the laws of ideal solutions, and have a
tendency toward ordering and not toward stratification as had been proposed [17, 18]. However,
our preliminary results require further investigation concerning the effect of prolonged annealing and also zone leveling on the thermodynamic properties of AlSb—GaSb solid solutions.
Further studies are being devoted to these topics.

TABLE 2. Values of the emf of Cell (2); Activity
of AlSb and GaSb; Excess Integral Gibbs Free
Energy of Formation of the AlSb and GaSb Solid
Solution at 855°K for Different Solid—Solution
Compositions

AlSb content	E, mV	a_{AlSb}	a_{GaSb}	$-\Delta G^{excess}$, kcal/mole
0.1	250	$3.7 \cdot 10^{-5}$	$7.2 \cdot 10^{-1}$	1.7
0.2	195	$3.5 \cdot 10^{-4}$	$4.8 \cdot 10^{-1}$	2.7
0.3	159	$1.5 \cdot 10^{-3}$	$3.0 \cdot 10^{-1}$	3.8
0.4	128	$5.4 \cdot 10^{-3}$	$1.5 \cdot 10^{-1}$	4.3
0.5	97,6	$1.9 \cdot 10^{-2}$	$5.45 \cdot 10^{-2}$	4.7
0.7	47,9	$1.4 \cdot 10^{-1}$	$3.0 \cdot 10^{-3}$	4.3
0.9	13,4	$5.8 \cdot 10^{-1}$	$1.7 \cdot 10^{-4}$	1.75

Literature Cited

1. B. F. Ormont, Proc. Fourth Conf. on Semiconducting Materials [in Russian], Akademkniga, Moscow (1961), p. 5.
2. S. A. Semenkovich, Dokl. Akad. Nauk SSSR, 158:442 (1964).
3. K. A. Sharifov and A. S. Abbasov, Dokl. Akad. Nauk SSSR, 157:430 (1964).
4. A. A. Vecher, V. A. Geiderikh, and Ya. I. Gerasimov, Zh. Fiz. Khim., 39:2145 (1965).
5. U. Piesbergen, Z. Naturforsch., 18a:141 (1963).
6. W. Kischio, J. Inorg. Nucl. Chem., 27:750 (1965).
7. M. Koch and K. S. Hinge, J. Chem. Phys., 35:451 (1961).
8. C. D. Thurmond, J. Phys. Chem. Solids, 26: 785 (1965).
9. S. I. Gadzhiev, Author's Abstract of Doctoral Thesis [in Russian], Baku (1962).
10. A. Schneider et al., Pure Appl. Chem., 2:13 (1961).
11. A. N. Krestovnikov et al., Handbook of Equilibrium Calculations for Metallurgical Reactions [in Russian], Metallurgizdat, Moscow (1962).
12. N. A. Goryunova, Chemistry of Diamond-like Semiconductors, Chapman and Hall, London (1965).
13. B. F. Ormont, Zh. Neorg. Khim., 3:1281 (1958).
14. J. P. Suchet, Chemical Physics of Semiconductors, Van Nostrand, London (1965).
15. I. I. Burdiyan and A. S. Borshchevskii, Zh. Tekh. Fiz., 28:2684 (1958).
16. J. F. Miller, H. L. Goering, and R. C. Himes, J. Electrochem. Soc., 107:527 (1960).
17. V. N. Romanenko and V. I. Ivanov-Omskii, Dokl. Akad. Nauk SSSR, 129:553 (1959).
18. G. V. Nikitina and V. N. Romanenko, Izv. Akad. Nauk SSSR, ser metallurgiya i gornoe delo, No. 6, p. 156 (1964).

EMF STUDY OF THE THERMODYNAMIC PROPERTIES
OF GALLIUM SELENIDE Ga_2Se_3*

A. S. Abbasov, K. N. Mamedov,
P. G. Rustamov, and P. K. Babaeva

An emf study was made of the thermodynamic functions relating to the formation (free energy, entropy, and enthalpy) of gallium selenide having the composition Ga_2Se_3. The experimental data were used to calculate the standard values of the free energy, entropy, and enthalpy of formation at 298°K.

According to Hansen and Anderko [1], gallium and selenium form three compounds: Ga_2Se_3, GaSe, and Ga_2Se. The existence of the compound Ga_2Se in thin layers had already been indicated [2].

The results of a recent investigation [3] by thermal and x-ray phase analysis, as well as a determination of the microstructure, enabled the authors to construct for the first time the phase diagram of the gallium–selenium system (Fig. 1), and to confirm the existence of the selenides cited above.

Later, the papers [4, 5] dealt with the phase diagram of the Ga–Se system and the polymorphism of Ga_2Se_3.

Gallium selenides exhibit semiconductor properties [6, 7]. Ga_2Se_3 belongs to the defect semiconductor group $A_2^{III}B_3^{VI}$ [8].

The thermodynamic properties of the gallium selenides have not been much studied. The purpose of the present investigation was to study the thermodynamic properties of Ga_2Se_3 by the emf method described in [9, 10].

The emf's of concentrated (relative to electrodes) electrochemical cells of the type

$$\ominus Ga_l \left| \begin{array}{c} \text{liquid} \\ \text{electrolyte} \end{array} + GaCl_3 \right| (Ga_xSe_{1-x})_s \oplus,$$

were studied, where x is molar fraction of gallium in the alloy.

The thermodynamic functions were calculated from the well-known relationships

$$\Delta G = -zFE, \tag{1}$$

---

* "Semiconductors," pp. 200-202 (see page 3).

where z is the charge of the potential-forming gallium ion $z_{Ga} = 3$ [11], F is the Faraday constant, equal to 23.062 cal/V · eq, and E represents emf;

$$\Delta S = zF \; \frac{dE}{dT} \; , \tag{2}$$

$$\Delta H = zF \left(T \; \frac{dE}{dT} - E \right) , \tag{3}$$

ΔG, ΔS, and ΔH being the changes in free energy, entropy, and enthalpy, respectively. The investigation was conducted over the temperature range 320–422°K.

The alloys were synthesized in evacuated (down to 10^{-3} mm Hg) and sealed quartz ampoules, followed by annealing (about 80 h). Gallium and selenium of 99.999% purity were used. The alloys prepared were subjected to x-ray phase and chemical analyses. The lattice parameter of $a = 5.43$ Å found for Ga_2Se_3 agrees with that given in the literature [12]. Anhydrous glycerol containing dissolved potassium chloride (4 wt.%) and added gallium chloride (0.1 wt.%), which was prepared by the method described in [13], was used as the electrolyte. Electrode alloys having the compositions 64.17, 75.04, 85.28, and 95.01 at.% Se, which are in the Ga_2Se_3–Se region, gave the same emf values, which corresponds to the formation of Ga_2Se_3 from gallium and selenium.

All the experimental emf data for the Ga_2Se_3–Se region were processed together by the method of least squares [14]. The equation $E = (0.721 - 0.102T \cdot 10^{-3}) \pm 0.012$ V was derived from this. Using this equation and Eqs. (1) − (3), we calculated the thermodynamic functions for the formation of Ga_2Se_3 at 320–422°K. The following results were obtained: $-\Delta G_{371}^0 = 18.9 \pm 0.3$ kcal/g-atom, $-\Delta S^0 = 2.8 \pm 0.5$ eu/g-atom, $-\Delta H^0 = 19.9 \pm 1.8$ kcal/g-atom.

The thermodynamic functions obtained from the experimental data for the formation of Ga_2Se_3 at 298°K, taking account of the heat of fusion and the entropy of the components [15], have the following values: $-\Delta G^0 = 19.1 \pm 0.3$ kcal/g-atom, $-\Delta S^0 = 1.1 \pm 0.5$ eu/g-atom, $-\Delta H^0 = 19.4 \pm 1.8$ and 21.0 ± 0.6 kcal/g-atom [15], $S^0 = 8.8 \pm 0.5$ eu/g-atom.

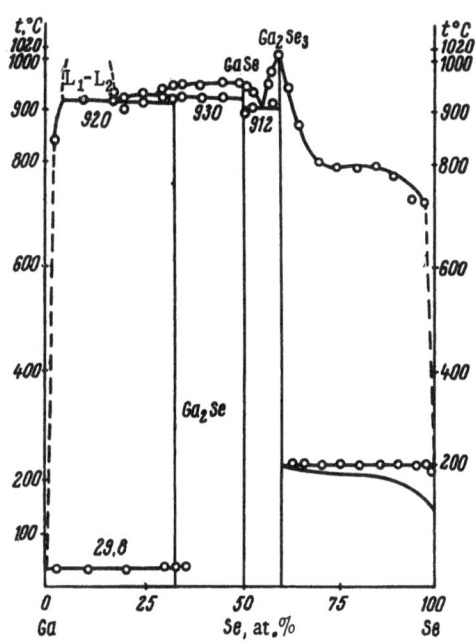

Fig. 1. Phase diagram of the gallium –
selenium system.

Literature Cited

1. M. Hansen and K. Anderko, Constitution of Binary Alloys, McGraw–Hill, New York (1958), p. 756.
2. K. Schubert and E. Dörre, Naturwiss., 40:604 (1953).
3. P. G. Rustamov, B. K. Babaeva, and N. P. Luzhnaya, Izv. Akad. Nauk SSSR, Neorg. Mater., 1:843 (1965).
4. L. S. Palatnik and E. K. Belova, Izv. Akad. Nauk SSSR, Neorg. Mater., 1:1883 (1965).
5. L. S. Palatnik and E. K. Belova, Izv. Akad. Nauk SSSR, Neorg. Mater., 2:770 (1966).
6. D. N. Nasledov and N. A. Fel'tinsh, Fiz. Tverd. Tela, 1:565 (1959).
7. E. Mooser and W. B. Pearson, J. Chem. Phys., 26:893 (1957).
8. N. A. Gorynova, Chemistry of Diamond–like Semiconductors, Chapman and Hall, London (1965).
9. Ya. I. Gerasimov, A. S. Abbasov, and A. V. Nikol'skaya, Dokl. Akad. Nauk SSSR, 147:835 (1962).
10. Ya. I. Gerasimov and A. V. Nikol'skaya, Problems in Semiconductor Metallurgy and Physics [in Russian], Izd. Akad. Nauk SSSR, Moscow (1961), p. 30.
11. H. A. Laitinen et al., Anal. Chem., 30:1266 (1958).
12. H. Hahn and W. Klingler, Z. Anorg. Allg. Chem., 259:135 (1949).
13. G. Brauer (ed.), Handbook of Preparative Inorganic Chemistry, Vols. 1 and 2, Academic Press, New York (1963, 1965).
14. V. V. Nalimov, Applications of Mathematical Statistics in Analysis of Materials [in Russian], Moscow (1960).
15. D. R. Stull and G. C. Sinke, Thermodynamic Properties of the Elements, American Chemical Society, Washington (1956).

THERMODYNAMIC PROPERTIES OF SOME SEMICONDUCTORS*

P. G. Maslov and Yu. P. Maslov

Sets of internally consistent equations, which are simple and convenient to use, have been derived by a new method to give explicit dependences of the thermodynamic properties of nineteen semiconductors: PbX, Y_2X_3 (X = S, Se, Te; Y = Bi, Sb) from 250°K to the melting point (with an accuracy of 0.5 to 3.0% and 1 to 4%), for semiconductors of the type InX (X = P, As, Sb) and GaX (X = As, Sb) from 30°K to the melting point (with an accuracy of 0.5 to 3%), for InSe from 30 to 400°K (with an accuracy of 0.1 to 0.8%), for BN, AlN, and SiC from 250 to 2900°K, 250 to 2700°K, and 250 to 3200°K, respectively, and for AlSb from 30°K to the melting point (with an accuracy of 0.5 to 3.0%).

A knowledge of the thermodynamic functions of semiconductors is extremely important in many respects. It is not by chance that over the last few decades more and more consideration has been given to this question and also to the connection between the thermodynamic, thermal, and electrical properties of semiconductors [1–8]. A number of extremely useful relationships have been discovered, associating the thermodynamic properties with the electrical and mechanical parameters of these materials [2]. These relationships have been analyzed in detail by Sirota [2]. Studies reported in [9, 10, 1–7] relate to similar topics.

Unfortunately, it is not possible to calculate theoretically the thermodynamic functions for the condensed phases of a material by the known methods of statistical thermodynamics; the empirical methods described in the literature by Kelley [11] and others are only approximate, although they have remained valuable up till this time.

The introduction of new methods [12-17], which have been semiempirically extended also to the liquid and solid phases [17], seems to have opened the way for a more effective study of the thermodynamic properties of solids, and particularly of semiconductors.

By combining the new methods [12–15] with the law of weighted means [17], we obtained sets of simple internally consistent equations which express the dependences of thermodynamic functions on temperature for nineteen important semiconductors: PbX, $Y_2 X_3$ (X = S, Se, Te; Y = Bi, Sb) from 250°K to the melting point, InX (X = P, As, Sb), GaX (X = As, Sb), and AlSb from 30°K to the melting point, SiC, AlN, and BN from 250°K to the melting point, and InSe from 30 to 400°K. In the literal notation the sets of equations are

$$C_p = c_1 + c_3 T + c_2 \log T, \tag{1}$$

$$H_T - H_0 = -\Phi_0 + h_1 T + 0.5 c_3 T^2 + c_2 T \log T, \tag{2}$$

---

* "Semiconductors," pp. 203-209 (see page 3).

TABLE 1. Coefficients for Eqs. (1)–(4) Representing the Dependences of Thermodynamic Properties of a Number of Semiconductors on Temperature (in cal·mole^{-1} and cal·mole^{-1}·deg^{-1})

Semi-conductor	c_1	c_2	$c_3 \cdot 10^7$	h_1	s_0	Φ_0	b_0	s_1	s_2	b_1	Temperature range, °K
PbS	-8.6215	8.06926	—	-12.1260	13.3122	-0.54	25.4382	-19.85169	9.29006	-27.92095	250—T$_{mp}$
PbSe	-8.4243	8.06926	—	-11.9288	14.6387	-0.55	26.5675	-19.39762	9.29006	-27.46688	250—T$_{mp}$
PbTe	-8.3415	8.06926	—	-11.8460	15.9360	-0.53	27.7829	-19.20697	9.29006	-27.27623	250—T$_{mp}$
InSe	-22.9468	17.16171	-258842	-30.4000	36.8216	-195.9	67.2216	-52.83684	19.75810	-69.99855	30—400
Bi$_2$S$_3$	10.3273	7.11403	—	7.2377	-63.2837		-70.5214	23.77943	8.19031	16.66540	250—T$_{mp}$
Bi$_2$Se$_3$	11.0373	7.21450	—	7.9041	-61.1371		-69.0412	25.41427	8.30598	18.19977	250—T$_{mp}$
Bi$_2$Te$_3$	11.5863	7.34060	—	8.3983	-58.1538		-66.5521	26.67838	8.45116	19.33778	250—T$_{mp}$
Sb$_2$S$_3$	8.4080	7.14203	—	5.3062	-63.0861		-68.3923	19.36009	8.22255	12.21806	250—T$_{mp}$
Sb$_2$Se$_3$	9.0410	7.25398	—	5.8906	-59.3417		-65.2323	20.81763	8.35144	13.56365	250—T$_{mp}$
Sb$_2$Te$_3$	9.7202	7.34728	—	6.5293	-57.6690		-64.1983	22.38154	8.45855	15.03426	250—T$_{mp}$
SiC	-48.6172	23.57687	-111734	-58.8565	118.1256	-1448.9	176.9821	-111.94497	27.14381	-135.52182	250—1000
	-7.4554	6.31166	—	-10.1960	1.5221	979	11.7181	-17.16569	7.26655	-23.47735	1000—3200
AlN	-32.4049	17.05451	-75038	-39.8116	71.4497	-551.4	111.2613	-74.61407	19.63448	-91.66858	250—1000
	27.2926	-6.28340	30725	30.0215	-109.6639	4555.3	-139.6854	62.84285	-7.23394	69.12625	1000—2700
InP	-12.4200	9.43159	—	-16.5161	19.3314	-95.3	35.8475	-28.59804	10.85850	-38.02963	30—160
	-6.0465	6.84883	—	9.0209	1.1873	212.1	10.2082	-13.92255	7.88499	-20.77138	140—1070
InAs	-14.6277	11.23286	—	-19.5061	22.3987	-112.6	41.9048	-33.68145	12.93228	-44.91431	30—160
	-1.1886	5.14225	—	-3.4218	-11.7206	227.5	-8.2988	-2.73685	5.92022	-7.87910	150—940
InSb	-14.3268	11.61553	—	-19.3714	21.2870	-103.2	40.6584	-32.98860	13.37284	-44.60413	30—150
	-0.7466	5.18221	—	-2.9972	-11.6502	256.0	-8.6530	-1.71911	5.96623	-6.90132	130—525,2
GaAs	-18.4678	12.70876	—	-23.9872	31.4513	106.2	55.4385	-42.52359	14.63147	-55.23235	20—200
	-3.0642	5.70418	—	-5.5415	-7.6285	-296.0	-2.0870	7.05557	6.56716	-12.75975	150—1238
GaSb	-23.6732	17.15830	-243658	-31.2496	39.3390	-195.9	70.5886	-54.50944	19.75418	-71.66774	40—300
	-1.0466	5.18221	—	-3.2972	-12.6122	378.0	-9.3150	-2.40995	5.96623	-7.59216	200—1052
AlSb	-8.2284	6.41478	192885	-11.0143	12.0564	-53.6	23.0707	-18.94655	7.38527	-25.36133	30—180
	-4.1238	6.17459	—	-6.8054	-4.6148	313.0	2.1906	9.49538	7.10874	-15.66997	180—1050
BN	-17.5137	8.86451	18492	-21.3636	40.4142	-386.1	61.7778	-40.32636	10.20552	-49.19087	250—1000
	2.6094	2.45046	9438	1.5452	-31.2251	2827.5	-32.7703	6.00831	2.82116	3.55785	1000—2900

TABLE 2. Comparison of the Thermodynamic Functions of Certain Semiconductors, Calculated from the Authors' Equations, with Data from Other Sources (in cal · mole^{-1} · deg^{-1})

T, °K	c_p			S			Φ^* potential		
	our eqs.	[19] data	Δ	our eqs.	[19] data	Δ	our eqs.	[19] data	Δ
				for AlN					
298.15	7.56	7.19	0.37	4.803	4.800	—0.003	1.684	1.701	—0.017
600	10.47	10.47	0.00	11.202	11.204	—0.002	4.964	4.966	—0 002
1000	11.52	11.42	0.10	16.814	16.814	0.000	8.663	8.662	—0.001
1500	11.94	12.02	—0.08	21.567	21.567	0.000	12.235	12.218	0.017
1800	12.37	12.31	0.06	23.781	23.784	—0 003	13 979	13.966	0.013
2000	12.70	12.69	0.01	25.100	25.100	0.000	15.026	15.014	0.012
2500	13.62	13.63	—0.01	28 030	28.032	—0.002	17.341	17.332	0.009
				for BN					
298.15	4.97	4 78	0.19	3.667	3.673	—0.006	1.525	1.523	0.002
400	6.29	6.34	—0.04	5.321	5.315	0.006	2.283	2.280	0.003
600	8.22	8.21	0.01	8.258	8.265	—0.007	3.797	3.795	0.000
800	9.70	9.67	0.03	10.834	10.831	0.003	5.241	5.240	0.001
1000	10.93	11.00	—0.07	13.134	13.134	0.000	6.593	6.593	0.000
1500	11.81	11.75	0.06	17.732	17.736	—0 004	9.581	9.583	—0.002
2000	12.59	12.50	0.09	21.238	21.219	0.019	12.072	12.073	—0.001
2300	12.88	12.80	0.08	23.026	22.997	0.029	13.388	13.384	0.004

$$S_T = s_0 + c_3 T + s_1 \log T + s_2 \log^2 T, \tag{3}$$

$$\Phi_T^* = b_0 + \Phi_0 T^{-1} + 0.5 c_3 T + b_1 \log T + s_2 \log^2 T. \tag{4}$$

The coefficients occurring in the equations are related to each other by expressions of the type [12–17]

$$h_1 = c_1 - 0.434295 c_2; \quad s_1 = 2.30258 c_1; \quad s_2 = 1.15129 c_2;$$
$$b_1 = s_1 - c_2; \quad b_0 = s_0 - h_1. \tag{5}$$

The coefficients s_0 and Φ_0 are found from the initial conditions; it is necessary to know either the geometry and molecular weights [13, 16], or the entropy and enthalpy or the entropy and the Φ^* potential at one temperature, for example, at 298.15°K, for the desired substance and the reference substance. In this case

$$s_0 = S_{T_0} - c_3 T_0 - s_1 \log T_0 - s_2 \log^2 T_0, \tag{6}$$

$$\Phi_0 = h_1 T_0 + 0.5 c_3 T_0^2 + c_2 T_0 \log T_0 - (H_{T_0} - H_0). \tag{7}$$

The internal consistency of the equations implies that they and the corresponding thermodynamic properties satisfy all the thermodynamic identities:

$$C_p = [\partial (H_T - H_0)/\partial T]_p = T(\partial S/\partial T)_p; \quad S = [\partial (\Phi^* T)/\partial T]_p. \tag{8}$$

For InSe, three values of C_p at three temperatures and values of S_T and enthalpy at 298.15°K, taken from [18], were used as the initial data. The accuracy of the equations for InSe is of the order of 0.1 to 0.8%. Corresponding expressions for SiC, BN, and AlN were obtained by using initial data from [19]; the accuracy is of the order of 0.1 to 0.5%. In deriving Eqs. (2)–(4) for $A^{III} B^V$-type semiconductors, the initial values were taken from [20]. The accuracy of Eqs. (1)–(4) is 0.5 to 3.0%; for C_p it is 0.5 to 4.0%.

TABLE 3. Comparison of the Thermodynamic Functions
of Certain Semiconductors, Calculated from the Authors'
Equations, with Data from Other Sources
(in cal $\cdot$ mole^{-1} $\cdot$ deg^{-1})

T, °K	c_p			s_T		
	from the authors' equations	data of [20]	Δ	from the authors' equations	data of [20]*	Δ
			for GaAs			
40	1 89	2.26	0.37	0.98	0.94	0.04
80	5.72	5.53	0.19	3.62	3.54	0.08
100	6.95	6.90	0.05	5.03	4.92	0.11
120	7.96	7.96	0.00	6.29	6.27	0.02
150	9.19	9.12	0.07	8.30	8.18	0.12
200	10.06	10.22	—0.16	10.91	10 97	—0.06
250	10.61	10.84	—0.23	13.22	13.33	—0.11
273,15	10.83	10.92	—0 09	14.16	14.16	0.00
			for AlSb			
40	2.82	2.82	0.00	1.43	1.43	0.00
80	5.52	5.42	0.10	4.40	4.23	0.17
100	6.53	6.53	0.00	5.63	5.56	0.07
120	7.42	7.46	—0.04	6.90	6.84	0.06
150	8.62	8.62	0.00	8.69	8.64	0.05
200	10.08	9.84	0.24	11.18	11.20	—0.02
250	10.68	10.68	0.00	13.49	13.49	0 00
273,15	10.92	10.92	0.00	14.45	14 45	0.00

*Entropy values, S_T, in this column were obtained from the specific heats
by graphical integration.

Unfortunately, for PbX- and Y_2X_3-type semiconductors (X = S, Se, Te; Y = Bi, Sb), data
in the literature are scarce, and also contradictory and not sufficiently accurate. According to
the data in [21], the entropy of PbS, PbSe, and PbTe at 298.15°K is, respectively, equal to 21.8 ±
0.6, 24.2 ± 1.5, and 26.1 ± 1.0 cal $\cdot$ mole^{-1} $\cdot$ deg^{-1}; these figures agree reasonably well among
themselves, but the accuracy of their determination is quite low. The corresponding values of
the entropy of these compounds quoted in [22] are inconsistent and therefore less reliable.
They are: S_T = 21.8, 26.9, and 17.6, respectively; the last two values were tentatively estimated
and do not agree with the first figure, as is easily seen by the law of weighted means. For this
reason, for PbX (X = S, Se, Te) the initial data were taken from [21]. By using the law of weighted means and the methods in [23], and using the specific heats C_p of PbS and PbO as the initial
data, we estimated the corresponding values for PbSe and PbTe. From similar considerations
the entropy and specific heat C_p were also estimated at 298.1°K for the Y_2X_3-type semiconductors (Y = Bi, Sb; X = S, Se, Te). Then, equations of type (1)–(4) were obtained for all the PbX,
Y_2X_3-type semiconductors (Y = Bi, Sb; X = S, Se, Te) by the method described in [12–17]. The
equations are valid in the temperature range 250°K to the melting point, but with considerably
less accuracy: 0.5 to 3.0% and 1 to 4%, respectively. However, using these equations, the entropy S_T of lead sulfide PbS at 298.15°K is 21.07 instead of the experimental value of 21.8 ± 0.6
cal $\cdot$ mole^{-1} $\cdot$ deg^{-1} [21]; 19 cal $\cdot$ mole^{-1} $\cdot$ deg^{-1} is obtained from Kelley's approximate equations
[11]. Similarly, the entropy of PbSe at the same temperature from the authors' equations is
23.52, and the experimental value is 24.2 ± 1.5 [21]; for PbTe, it is 25.3 as opposed to 26.6 ± 1.0
cal $\cdot$ mole^{-1} $\cdot$ deg^{-1} in [21]. Also, the specific heat of Bi_2Se_3 at 80°K from the authors' equations
is 24.8, whereas experiment gives 23.5 cal $\cdot$ mole^{-1} $\cdot$ deg^{-1}; at the same temperature for Bi_2Te_3
from the equations we have 25.6 as opposed to 24.76 by experiment. At 300°K from the equa-
tions, the specific heat C_p of Bi_2Te_3 is equal to 29.75; the experimental value is 29.75 cal $\cdot$
mole^{-1} $\cdot$ deg^{-1}. As can be seen, there is reasonable agreement with experimental values even
well outside the temperature limits for which the equations were derived.

However, under these conditions, the proposed relationships should be used only for purposes of estimation until more accurate data are available. The equations may nevertheless still be useful in this case. Also, they can always be made more accurate when more accurate initial values become available.

Calculated coefficients are shown in Table 1 for all nineteen semiconductors. Comparative material, given in Tables 2 and 3, shows the satisfactory agreement between the calculated values, determined from the authors' equations, and the results from other sources.

Literature Cited

1. A. F. Ioffe, Physics of Semiconductors, Infosearch, London (1960).
2. N. N. Sirota, in: Chemical Bonds in Semiconductors and Solids (ed. by N. N. Sirota), Consultants Bureau, New York (1967), p. 7.
3. N. N. Sirota, Abstracts of Papers, Eighth Mendeleev Meeting, Metals Section [in Russian], Izd. Akad. Nauk SSSR, Moscow (1959).
4. A. S. Borshchevskii, N. A. Goryunova, and N. K. Takhtareva, Zh. Tekh. Fiz., 27:1408 (1957).
5. N. N. Sirota, Dokl. Akad. Nauk SSSR, 47:40 (1945).
6. N. F. Mott and H. Jones, The Theory of the Properties of Metals and Alloys, Oxford University Press, Oxford (1945).
7. N. N. Sirota and M. I. Danil'kevich, Dokl. Akad. Nauk Belorus. SSR, 8:369 (1964).
8. N. N. Sirota, in: Chemical Bonds in Semiconductors and Thermodynamics (ed. by N. N. Sirota), Consultants Bureau, New York (1968), p. 1.
9. S. A. Semenkovich, Zh. Fiz. Khim., 39:2232 (1965).
10. P. G. Maslov, Zh. Fiz. Khim., 25:814 (1951).
11. K. K. Kelley, U. S. Bur. of Mines Bull. 584, Washington (1960).
12. P. G. Maslov, A. A. Antonov, and Yu. P. Maslov, Zh. Fiz. Khim., 40:362 (1966).
13. P. G. Maslov, A. S. Stepanov, and Yu. P. Maslov, Zh. Obshch. Khim., 36:1355 (1966).
14. A. A. Antonov, Thesis [in Russian], A. I. Gertsen Leningrad State Pedagogical Institute (1967).
15. P. G. Maslov, A. A. Antonov, and Yu. P. Maslov, Teplofiz. Vys. Temp., 5:278 (1967).
16. Yu. P. Maslov, Zh. Fiz. Khim., 35:974 (1961).
17. P. G. Maslov and Yu. P. Maslov, Elektron. Tech., Ser. 10, No. 3, p. 70 (1967).
18. K. K. Mamedov, I. G. Kerimov, V. N. Kostryukov, and G. D. Guseinov, in: Chemical Bonds in Semiconductors and Thermodynamics (ed. by N. N. Sirota), Consultants Bureau, New York (1968), p. 132.
19. V. P. Glushko (editor), Thermodynamic Properties of Individual Substances [in Russian], Vols. 1 and 2, Izd. Akad. Nauk SSSR, Moscow (1962).
20. U. Piesbergen, Z. Naturforsch., 18:141 (1963).
21. O. Kubaschewski and E. Ll. Evans, Metallurgical Thermochemistry, Butterworth-Springer, London (1951).
22. W. M. Latimer, The Oxidation States of the Elements and Their Potentials in Aqueous Solutions, Prentice-Hall, New York (1952).
23. M. Kh. Karapet'yants, Methods of Comparative Calculation of Physical and Chemical Properties [in Russian], Izd. "Nauka," Moscow (1965).

KINETICS OF THE ETCHING OF
A^{IV} AND $A^{III}B^{V}$ SEMICONDUCTORS*

G. M. Orlova

The results are presented of an investigation into the rate of dissolution of germanium and $A^{III}B^{V}$ compounds in acidic solutions of iron (III) chloride. The kinetics of dissolution agrees with ideas on the structures of these compounds and the nature of the chemical bonds in them.

The chemical stability of a semiconductor is determined mainly by its chemical composition and the nature of the chemical bonds predominating in it and may be characterized by its rate of dissolution in a given etchant. Consequently, a correlation was to be expected between the kinetics of the dissolution process and the nature of the chemical bonds between the atoms in the compound. The steady-state rate of dissolution of a solid is determined by the activation energy of the limiting process.

Under otherwise equal conditions, the covalently rigidly bound particles in a solid should expend more energy in overcoming bond forces on passing into solution, and their dissolution should be accompanied by higher activation energies, than in the dissolution of ionic solids. It is known that the rate of dissolution of solids containing a covalent bond is determined by the rate of chemical reaction on the surface of these substances with an activation energy ≥ 10 kcal/mole.

It was of interest to study the effect of the ionic component and of metallization of the chemical bonds on the kinetics of etching single-crystal germanium and compounds of the type $A^{III}B^{V}$.

The chemical etching of semiconductors is not determined only by the properties of the solid phase; no less important are such factors as the composition of the solution, its oxidizing power, rate of dissolution of the oxidation products, etc. The dissolution kinetics of single crystals of Ge, GaAs, InAs, GaSb, and InSb was studied in solutions containing various oxidizing agents (H_2O_2, I_2, HNO_3, $K_3[Fe(CN)_6]$, $FeCl_3$, etc.).

In the present paper, the results are presented of an investigation into the etching of germanium and $A^{III}B^{V}$ compounds in acidic solutions of iron (III) chloride. The rate of dissolution (W) was determined with an accuracy of 10% by a well-known method [1]. The rate of dissolution was studied as a function of the stirring of the solution, temperature, and the concentration of the oxidizing agent in solution. The state of the surfaces of the specimens was checked with an MIM−7 instrument.

---

* "Semiconductors," pp. 210−216 (see page 3).

TABLE 1. Kinetic Data for the Dissolution of Single-Crystal
Ge, GaAs, and GaSb in Hydrochloric Acid Solutions of Iron (III)
Chloride

Substance	$FeCl_3$ concn., N	HCl concn., N	Stirring, γ	E_a, kcal per mole	C_e, mole $\cdot$ cm$^{-2} \cdot$ sec^{-1}	n_e, mole per cm^2
Ge	1.7	7.0	$\mp$	17.2	$1 \cdot 10^{27}$	$1 \cdot 10^{15}$
GaAs	1.6	2.2	$\mp$	12.4	$4 \cdot 10^{24}$	$4 \cdot 10^{12}$
GaAs	1 5	6.7	$\mp$	12.4	$1 \cdot 10^{24}$	$1 \cdot 10^{12}$
GaSb	1.6	2.2	$-$	3.0	$1 \cdot 10^{19}$	$1 \cdot 10^7$
			$+$	5.0	$5 \cdot 10^{20}$	$5 \cdot 10^8$

The quantities E_a and C_e in the equation $W = C_e e^{-E_a/RT}$, found from the temperature dependence of the rate of dissolution, and n_e, the surface concentration of active particles transferring into solution in unit time ($n_e = C_e/\nu$, where $\nu \approx 10^{12}-10^{13}$ sec^{-1} is the frequency of the thermal vibrations of the solid particles) as well as data on the variation in the rate of dissolution with stirring of the solution, enable the nature of the dissolution process to be judged.

The data in Table 1 show that the rate of dissolution of germanium in hydrochloric acid solutions of iron (III) chloride is determined by the rate of the heterogeneous chemical reaction having an activation energy (E_a) of approximately 17 kcal/mole. To within an order of magnitude, all the surface atoms of germanium (n_s) transfer into solution. In the solutions studied $n_s \approx 10^{15}$ at./cm^2, i.e., $n_e \approx n_s$.

The rate of dissolution of gallium arsenide is also determined principally by the rate of the chemical reaction on the surface of a single crystal, having an activation energy of approximately 12 kcal/mole. However, the value of n_e is found to be low compared with n_s. In contrast to germanium and gallium arsenide, gallium antimonide dissolves with lower activation energies, and to a greater extent in a stirred solution compared with a static solution. Moreover, GaSb is characterized by $n_e \ll n_s$. The data presented testifies to a limitation of the dissolution of GaSb by diffusion in solution.

Confirmation of the diffusional nature of the dissolution of gallium and indium antimonides was obtained from a rotating-disk study of their dissolution. It was found that the rate of dissolution of these antimonides increases with stirring of the solution, the values of the temperature coefficients (1.2 ± 0.1) correspond to the temperature coefficient of the diffusion process, the rate of dissolution increases in direct proportion to the increase in concentration of the oxidizing agent in solution and is virtually independent of the nature of the solid phase and the microrelief of the surface, good agreement being observed between the experimentally determined rates of dissolution and those calculated from the Levich equation for the rate of diffusion of neutral particles to the surface of a rotating disk. Thus, it was shown unequivocally that the rate of dissolution of gallium and indium antimonides is determined by the diffusion of iron (III) chloride to the surface of the single crystals. Taking into account the information [2–4] available on the structure of single crystals of $A^{III}B^V$ compounds oriented in the [111] crystallographic direction, it seemed advisable to study the dissolution kinetics of the {111} and {$\bar{1}\bar{1}\bar{1}$} surfaces of gallium and indium arsenides.

As may be seen from Fig. 1, on passing from germanium to gallium and indium arsenides, the rate of dissolution gradually increases, the B (or "arsenic") surfaces of GaAs and InAs dissolving more rapidly than the A (or "gallium" or "indium," respectively) surfaces.

The data in Table 2 show that the dissolution of Ge and GaAs in perchloric acid solutions of iron (III) chloride, as in hydrochloric acid solutions, takes place in the kinetic region, i.e., it is determined by the rate of the heterogeneous chemical reaction.

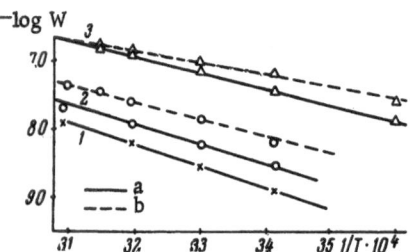

Fig. 1. Log W versus 1/T relationship for the etching of single crystals of Ge (1), GaAs (2), and InAs (3), oriented in the [111] direction, in a 2.0 N perchloric acid solution of 1.5 N iron (III) chloride. a) A surface; b) B surface.

The dissolution of InAs takes place in the transition region, the rate of dissolution being higher and the corresponding characteristic quantities (E_a, C_e, and n_e) being lower than for gallium arsenide. Furthermore, lower values of E_a, C_e, and n_e are obtained for the dissolution of the B surfaces of GaAs and InAs compared with the A surfaces, which testifies to diffusion having some effect on the rate of dissolution of the B surfaces. Etch patterns characteristic of {111} surfaces were observed only on the A surfaces. With an appreciable deviation from the orientation (up to 10°), distortion of the etch patterns was observed (Fig. 2).

Table 3 shows the results of studying the effect of the type and magnitude of the conductivity of the specimens and the effect of light from a steady source on the dissolution kinetics of single-crystal gallium arsenide oriented along the [111] direction.

As may be seen from Table 3, the type and magnitude of the conductivity of the specimens have no marked effect on the nature of the dissolution of GaAs. The rate of dissolution was approximately the same for both n-type GaAs and p-type GaAs, but was always higher for the B surfaces. A change in the type and magnitude of the conductivity did not affect the activation energies either. Under the action of light, a reduction was observed in the activation energy, this being particularly marked in the dissolution of the A surfaces of low-resistance specimens of n-type GaAs, where the activation energy was lowered by 4−5 kcal/mole on illumination. The activation energies and other characteristic quantities in the dissolution of the B surfaces of specimens of GaAs underwent little change under the action of light.

The etching kinetics of germanium and $A^{III}B^V$ compounds in acidic solutions of iron (III) chloride agree with ideas on the structure of these compounds and the nature of the chemical bonds in them.

According to current ideas, the atoms in germanium and $A^{III}B^V$ compounds are joined by chemical bonds having an appreciable covalent component. The predominantly covalent nature of the bonds in germanium and its closest electronic analog, viz., gallium arsenide, causes their rate of dissolution to be limited by the rate of the heterogeneous reaction.

TABLE 2. Kinetic Data for the Dissolution of the A and B Surfaces of Single-Crystal Ge, GaAs, and InAs, Oriented along the [111] Direction in 2.0 N Perchloric Acid Solutions of Iron (III) Chloride

Substance	Type of conduction	Resistivity, $\Omega \cdot cm$	FeCl$_3$ concn., N	A surface			B surface		
				E_a, kcal/mole	C_e, mole·cm^{-2}·sec^{-1}	n_e, mole/cm^2	E_a, kcal/mole	C_e, mole·cm^{-2}·sec^{-1}	n_e, mole/cm^2
Ge	n	30	1.5	14 0	$2 \cdot 10^{25}$	$2 \cdot 10^{13}$	14.0	$2 \cdot 10^{25}$	$2 \cdot 10^{13}$
GaAS	n	$2 \cdot 10^{-2}$	1,3	13.2	$1 \cdot 10^{25}$	$1 \cdot 10^{13}$	10.0	$2 \cdot 10^{23}$	$2 \cdot 10^{11}$
InAS	n	$2 \cdot 10^{-3}$	1.5	10,0	$5 \cdot 10^{23}$	$5 \cdot 10^{11}$	6.6	$3 \cdot 10^{21}$	$3 \cdot 10^{9}$

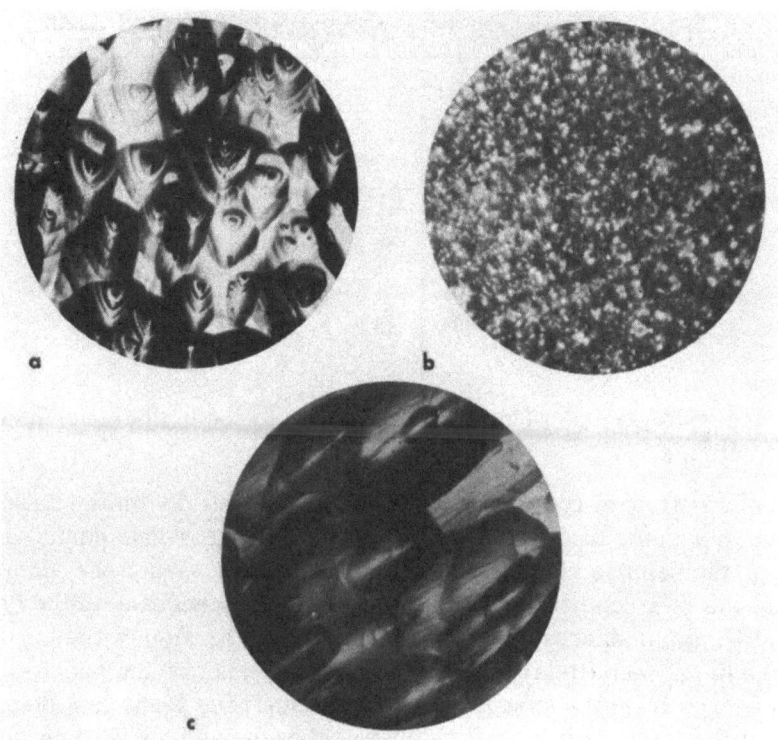

Fig. 2. Photomicrographs of the opposite surfaces of a GaAs single crystal oriented in the [111] direction after etching in a 1.5 N solution of iron (III) chloride in 2.0 N perchloric acid. a) A surfaces ($\times$120); b) B surfaces ($\times$340); c) A surfaces of a single crystal with an appreciable deviation from orientation in the [111] direction ($\times$70).

Data on the dissolution kinetics of GaAs and the limitation of the process by the rate of the chemical reaction on its surface poses the problem of the atomic-molecular mechanism of the dissolution of gallium arsenide in acidic solutions of iron (III) chloride.

From these data, in accordance with Myuller's concepts [5], it can be supposed that the rate-determining stage in the dissolution of a single crystal of gallium arsenide oriented along the [111] direction will be, for the A surface, the rupture of one of the three bonds between the surface atoms of gallium and the underlying arsenic atoms with the simultaneous rearrangement of the valence electrons at the surface atoms. The transfer of the arsenic atoms into solution, following the gallium atoms, should not limit the rate of dissolution of a single crystal, because the arsenic atoms are connected with the lattice by only one bond [4]. On the B surface, the rate of dissolution will be limited by the rupture of one of the three bonds between the surface arsenic atoms and the underlying gallium atoms, but in this case bond rupture is facilitated owing to the resulting excess negative charge at the surface arsenic atoms.

The ionization of the atoms in $A^{III}B^{V}$ compounds caused by the displacement of the electron cloud from the group III atoms to the group V atoms [6] leads to an increase in the rate of dissolution and a marked effect of diffusion on the rate of dissolution of the B surfaces of single crystals of arsenides. The effective charges of the ions decrease gradually in the order aluminum–gallium–indium and are higher for the arsenides than for the antimonides [6]. The more marked difference in the rates of dissolution of the A and B surfaces of gallium arsenide com-

TABLE 3. Activation Energies for the Dissolution of GaAs
in a 1.3 N Solution of $FeCl_3$ in 2.0 N $HClO_4$

Type of conduction	Resistivity, $\Omega \cdot cm$	Carrier density, cm^{-3}	E_a, kcal/mole			
			A surface		B surface	
			dark	illuminated	dark	illuminated
n	$1 \cdot 10^{-3}$	$3 \cdot 10^{18}$	13.2	8.2	10.7	10.3
n	$1 \cdot 10^{-2}$	$1 \cdot 10^{17}$	13.5	8.7	10.9	10.7
n	$2 \cdot 10^{-2}$	$8 \cdot 10^{16}$	13.2	8.9	10.9	9.2
n	$3 \cdot 10^{3}$	$1 \cdot 10^{13}$	13.0	11.1	11.1	10.7
p	$1 \cdot 10^{-2}$	$4 \cdot 10^{18}$	13.2	9.6	10.7	9.6

pared with indium arsenide (Fig. 1) would seem to be due to the high degree of ionization of the atoms in gallium arsenide.

The absence of an effect of type and magnitude of conductivity on the rate of dissolution for GaAs enables us to assume that the reduction of trivalent iron ions during dissolution takes place mainly through the valence electrons of gallium arsenide. When the atomic numbers of the component elements of a compound increase, the electron densities in the "electron bridges" between nearest unlike atoms decrease, i.e., a marked metallization of the bonds is observed [6, 7]. The increase in the metallization of the chemical bonds and the decrease in the interatomic interaction energy from the arsenides to the antimonides leads to a decrease in the rate of dissolution of the latter in iron (III) chloride solutions and determines the course of the dissolution process in the diffusion region.

Literature Cited

1. R. L. Myuller, T. P. Markova, and S. M. Repinskii, Vestnik Leningrad. Univ., Fiz. Khim., No. 16(3), p. 106 (1958).
2. E. P. Warekois and P. H. Metzger, J. Appl. Phys., 30:960 (1959).
3. J. G. White and W. C. Roth, J. Appl. Phys., 30:946 (1959).
4. H. C. Gatos et al., The Surface Chemistry of Metals and Semiconductors, Wiley, New York (1960).
5. R. L. Myuller, A. V. Danilov, T. P. Markova, V. N. Mel'nikov, A. B. Nikol'skii, and S. M. Repinskii, Vestnik Leningrad. Univ., Fiz. Khim., No. 4(1), p. 80 (1960).
6. N. N. Sirota and E. M. Gololobov, Dokl. Akad. Nauk SSSR, 156:1075 (1964).
7. N. A. Goryunova, Chemistry of Diamond-like Semiconductors, Chapman and Hall, London (1965).